Emerging Functions of Nano-Organized Polysaccharides

Emerging Functions of Nano-Organized Polysaccharides

Editor

Takuya Kitaoka

MDPI • Basel • Beijing • Wuhan • Barcelona • Belgrade • Manchester • Tokyo • Cluj • Tianjin

Editor
Takuya Kitaoka
Agro-Environmental Sciences
Kyushu University
Fukuoka
Japan

Editorial Office
MDPI
St. Alban-Anlage 66
4052 Basel, Switzerland

This is a reprint of articles from the Special Issue published online in the open access journal *Nanomaterials* (ISSN 2079-4991) (available at: www.mdpi.com/journal/nanomaterials/special_issues/nano_polysaccharides).

For citation purposes, cite each article independently as indicated on the article page online and as indicated below:

LastName, A.A.; LastName, B.B.; LastName, C.C. Article Title. *Journal Name* **Year**, *Volume Number*, Page Range.

ISBN 978-3-0365-4044-3 (Hbk)
ISBN 978-3-0365-4043-6 (PDF)

Contents

About the Editor

Takuya Kitaoka

Takuya Kitaoka is a professor at the Department of Agro-Environmental Sciences, Kyushu University, Japan, in charge of nanomaterials chemistry and sustainable bioresources science. He graduated from The University of Tokyo, Japan, in 1993. He received his M.Sc. (Forest Products Science) in 1995 and his Ph.D. (Agricultural and Life Science) in 2000, both from The University of Tokyo, Japan. He started his career at the National Printing Bureau (Japanese banknote manufacturer) in 1995. In 1998, he changed his occupation to Kyushu University, as an Assistant Professor. He promoted an Associate Professor in 2003 and the current position in 2013. He was engaged in paper and cellulose chemistry, especially for paper sizing, wet-end process chemistry, and functional paper composites. Recently, he has expanded his research into the advanced fields of biomaterials and catalysts, e.g., nonaqueous biocatalysis, biomedical biomaterials, organocatalytic nanomaterials and organic-inorganic hybrid nanomaterials. He has a lot of academic achievement as follows: 127 publications in international journals, 21 book chapters, 35 review articles, 27 patent applications and 21 invited/keynote speakers in international conferences. He was awarded "The Young Scientists' Prize" by the Minister of Education, Culture, Sports, Science and Technology, Japan in 2007, "JSPS PRIZE" by the Japan Society for the Promotion of Science in 2011, "The Cellulose Society of Japan Award" by the Cellulose Society of Japan in 2013, and "Fiber Science and Technology Award" by the Society of Fiber Science and Technology, Japan in 2014.

Preface to "Emerging Functions of Nano-Organized Polysaccharides"

Sustainable biomass has strongly attracted much attention from the international community, as well as academic and industrial circles, since further utilization of renewable bioresources will provide a key clue to solve various environmental issues for energy and materials production in the world.

Natural polysaccharides, such as cellulose and chitin, are the major biomasses in large quantity. More to the point, they possess unique hierarchical nanoarchitectures, e.g., crystalline, fibrous, and needle-like structures, in which each macromolecular component assembles in a regular and organized manner during bio-synthesis and/or physicochemical processing. Such organized nanoarchitectures can never be artificially reconstructed. Among the various nano-organized polysaccharides, nanocellulose obtained from plants and bacteria is the most promising natural nanomaterial in practical applications due to its high aspect ratio, high elastic modulus, high transparency, low thermal expansion coefficient, and other fascinating properties. Chitin nanofibers from crabs and shrimps are also expected to produce advanced nanomaterials for cosmetics and biomedical applications. Starch, xylan, and other polysaccharides have recently been studied for novel applications as nanomaterials, instead of traditional bulk uses. Research on and development of natural polysaccharides are classified into two categories: (1) greener alternatives to existing products from ecological and sustainability viewpoints and (2) emerging functional nanomaterials from scientific encounters with the unknown. Both approaches are important for advancing the utilization of natural polysaccharides.

This Special Issue book, entitled "Emerging Functions of Nano-Organized Polysaccharides", aims to showcase the current challenges involved with conceptual and functional designs of nano-polysaccharide materials for a diverse range of future applications. The unexpected novel functions arising from the inherent nanoarchitectures of natural nano-organized polysaccharides will provide new insights into polysaccharide-driven nanomaterial chemistry and engineering. This book comprises one editorial, one review paper and ten research articles, all of which will inspire readers to learn more for various natural polysaccharides and unique strategies for functional materials design. In this context, this international book will give the active readers perspectives different before, to open up new avenues for various emerging functions of natural nano-polysaccharides.

The Editor expresses the greatest appreciation to all the contributors for their excellent works, and the editorial board members for their support throughout the process of setting up this Special Issue book.

I hope you enjoy reading this reprint.

Takuya Kitaoka
Editor

nanomaterials

Editorial

Emerging Functions of Nano-Organized Polysaccharides

Takuya Kitaoka 

Department of Agro-Environmental Sciences, Graduate School of Bioresource and Bioenvironmental Sciences, Kyushu University, Fukuoka 819-0395, Japan; tkitaoka@agr.kyushu-u.ac.jp; Tel.: +81-92-802-4665

Citation: Kitaoka, T. Emerging Functions of Nano-Organized Polysaccharides. *Nanomaterials* **2022**, *12*, 1277. https://doi.org/10.3390/nano12081277

Received: 1 April 2022
Accepted: 7 April 2022
Published: 8 April 2022

Publisher's Note: MDPI stays neutral with regard to jurisdictional claims in published maps and institutional affiliations.

Natural polysaccharides, such as cellulose and chitin, possess unique hierarchical nanoarchitectures, e.g., crystalline, fibrous, and needle-like structures, in which each macromolecular component assembles in a regular and organized manner during biosynthesis and/or physicochemical processing. Among the various nano-organized polysaccharides, nanocellulose obtained from plants and bacteria is the most promising natural nanomaterial in practical applications due to its high aspect ratio, high elastic modulus, high transparency, low thermal expansion coefficient, and other fascinating properties. Chitin nanofibers from crabs and shrimps are also expected to produce advanced nanomaterials for cosmetics and biomedical applications. Starch, xylan, and other polysaccharides have recently been studied for novel applications as nanomaterials. Research on and development of natural polysaccharides are classified into two categories: (1) greener alternatives to existing products from ecological and sustainability viewpoints and (2) emerging functional nanomaterials from scientific encounters with the unknown. Both approaches are important for advancing the utilization of natural polysaccharides.

The aim of this Special Issue, entitled *Emerging Functions of Nano-Organized Polysaccharides*, is to showcase the current challenges involved with new conceptual and functional designs of nano-polysaccharide materials for a diverse range of future applications. The unexpected new functions arising from the inherent nanoarchitectures of natural nano-organized polysaccharides will provide new insights into polysaccharide-driven nanomaterial chemistry and engineering. This Special Issue comprises one review paper and ten research articles. The combination of various natural polysaccharides and unique strategies for functional material design will open up new avenues for the emerging functions of natural nano-polysaccharides.

First of all, the review article provided by Miyagi and Teramoto comprehensively covers the recent advances in the construction of functional materials in various material forms from cellulosic cholesteric liquid crystals [1]. The spontaneous and controllable self-assembling features of cellulosic nanomaterials are the most fascinating phenomenon for materials design, and this review provides information on both the fundamentals and the structure-driven functions of cellulosic cholesteric nanomaterials in various forms.

Strong, tough, and high-performance nanocellulose composites are typical materials that draw public attention. Sakuma et al. successfully fabricated millimeter-thick, transparent cellulose nanofiber (CNF)/polymer composites using CNF xerogels with high porosity as a template [2]. The as-designed composites demonstrate high flexural modulus, low thermal expansion, and good flame-retardant properties. Beaumont et al. highlight the solvent-free sol/gel process for the facile preparation of mechanically robust CNF/silica nanocomposites in combination with methylcellulose and starch [3]. Strong and shapable 3D CNF/silica hydrogels can act as adsorbers for heavy metals and dyes. Zhu et al., developed 3D nanocarbon materials via the polydopamine doping and pyrolysis of CNF paper [4]. This work achieved a high yield of 3D porous nanocarbons with good specific capacitance up to 200 F g^{-1}. Tsuneyasu et al. exhibit the successful enhancement of luminance in powder electroluminescent devices using smooth and transparent CNF films [5]. The optical and morphological properties of CNF substrates make a great contribution to increasing both the current density and luminance of the devices. Fukuda et al. introduce the enzymatic preparation of spherical microparticles from artificial lignin and CNF for cosmetic applications [6]. Lignin is also an abundant, renewable, and biodegradable biopolymer, comparable to natural polysaccharides, and the combination of lignin and CNF will attract much attention in a carbon-neutral eco-society. Studies like those can expand the possibility of creating high-value nanomaterials with nano-polysaccharides like CNF.

Apart from nanocellulose, other polysaccharides are also promising for the design of advanced functional materials. Xylan is a highly abundant, plant-based form of biomass and is expected to find further uses. Wang and Xiang successfully prepared a highly stable Pickering emulsion with xylan hydrate nanocrystals with a uniform size [7]. This is a pioneering work in the use of xylan-derived nanomaterials as a solid surfactant emulsifier. Chitin is the second most abundant polysaccharide after cellulose and has unique hierarchical nanoarchitectures similar to those of nanocellulose. Wang et al. report the thermal conductivity of chitin and deacetylated-chitin nanofiber films and reveal the relationship between the in-plane thermal conductivity and surface chemistry [8]. A computational and theoretical study on non-covalent interactions with nano-polysaccharides is of great significance to advancing the materials science of natural biomaterials. Yui et al. provide computational predictions for the molecular and crystal structures of a chitosan–zinc chloride complex [9]. Chitosan chains of antiparallel polarity form zigzag-shaped chain sheets, and the coordination of Zn ions with chitosan has been proposed. In the days ahead, the digital transformation of materials science will become more and more important for the unexpected discovery of unknown functions of nano-organized polysaccharides.

The biomedical applications of natural nano-polysaccharides have recently aroused great interest in tissue engineering and various therapeutics. Noda et al. report on the unique combination of surface-carboxylated nanocellulose and surface-deacetylated chitin nanofibers to promote the adhesion, migration, and proliferation of fibroblast cells [10]. Bioinert CNF can be designed to prepare bioadaptive materials by surface carboxylation, and cellular migration can be controlled using special nano-polysaccharides with a core–shell structure of chitin and chitosan, respectively. Moreover, bacterial nanocellulose is also a fascinating nanomaterial, as well as plant-derived nanocellulose. Ando et al. emphasize the feasibility of nanofibrillated bacterial cellulose as a carrier of doxorubicin to promote therapeutic outcomes in gastric cancer [11]. Bacterial cellulose is very effective for the drug loading and release of various anticancer agents, making it a promising nanomaterial for advanced drug delivery systems in cancer therapy.

In summary, pioneering work on the extraordinary and emerging functions of nano-organized polysaccharides, such as nanocellulose, chitin/chitosan nanofibers, and other nano-polysaccharides, will signal a new trend in the research and development of natural nanomaterials as we move towards achieving the UN's Sustainable Development Goals. All the papers published in this Special Issue will present readers with new concepts and ideas for the next-generation material design of nano-organized polysaccharides. Last but certainly not least, I would like to express my great appreciation to all the authors for their excellent work, the reviewers for their constructive contributions, and the editorial board members for their support throughout the process of setting up this Special Issue.

Funding: This research was funded by the JST-Mirai Program (grant no. JPMJMI21EC) from the Japan Science and Technology Agency.

Conflicts of Interest: The author declares no conflict of interest.

References

1. Miyagi, K.; Teramoto, Y. Construction of Functional Materials in Various Material Forms from Cellulosic Cholesteric Liquid Crystals. *Nanomaterials* **2021**, *11*, 2969. [CrossRef]
2. Sakuma, W.; Fujisawa, S.; Berglund, L.A.; Saito, T. Nanocellulose Xerogel as Template for Transparent, Thick, Flame-Retardant Polymer Nanocomposites. *Nanomaterials* **2021**, *11*, 3032. [CrossRef] [PubMed]

3.   Beaumont, M.; Jahn, E.; Mautner, A.; Veigel, S.; Böhmdorfer, S.; Potthast, A.; Gindl-Altmutter, W.; Rosenau, T. Facile Preparation of Mechanically Robust and Functional Silica/Cellulose Nanofiber Gels Reinforced with Soluble Polysaccharides. *Nanomaterials* **2022**, *12*, 895. [CrossRef] [PubMed]

4.   Zhu, L.; Uetani, K.; Nogi, M.; Koga, H. Polydopamine Doping and Pyrolysis of Cellulose Nanofiber Paper for Fabrication of Three-Dimensional Nanocarbon with Improved Yield and Capacitive Performances. *Nanomaterials* **2021**, *11*, 3249. [CrossRef] [PubMed]

5.   Tsuneyasu, S.; Watanabe, R.; Takeda, N.; Uetani, K.; Izakura, S.; Kasuya, K.; Takahashi, K.; Satoh, T. Enhancement of Luminance in Powder Electroluminescent Devices by Substrates of Smooth and Transparent Cellulose Nanofiber Films. *Nanomaterials* **2021**, *11*, 697. [CrossRef] [PubMed]

6.   Fukuda, N.; Hatakeyama, M.; Kitaoka, T. Enzymatic Preparation and Characterization of Spherical Microparticles Composed of Artificial Lignin and TEMPO-oxidized Cellulose Nanofiber. *Nanomaterials* **2021**, *11*, 917. [CrossRef] [PubMed]

7.   Wang, S.; Xiang, Z. Highly Stable Pickering Emulsions with Xylan Hydrate Nanocrystals. *Nanomaterials* **2021**, *11*, 2558. [CrossRef] [PubMed]

8.   Wang, J.; Kasuya, K.; Koga, H.; Nogi, M.; Uetani, K. Thermal Conductivity Analysis of Chitin and Deacetylated-Chitin Nanofiber Films under Dry Conditions. *Nanomaterials* **2021**, *11*, 658. [CrossRef] [PubMed]

9.   Yui, T.; Uto, T.; Ogawa, K. Molecular and Crystal Structure of a Chitosan–Zinc Chloride Complex. *Nanomaterials* **2021**, *11*, 1407. [CrossRef] [PubMed]

10.   Noda, T.; Hatakeyama, M.; Kitaoka, T. Combination of Polysaccharide Nanofibers Derived from Cellulose and Chitin Promotes the Adhesion, Migration and Proliferation of Mouse Fibroblast Cells. *Nanomaterials* **2022**, *12*, 402. [CrossRef] [PubMed]

11.   Ando, H.; Mochizuki, T.; Lila, A.S.A.; Akagi, S.; Tajima, K.; Fujita, K.; Shimizu, T.; Ishima, Y.; Matsushima, T.; Kusano, T.; et al. Doxorubicin Embedded into Nanofibrillated Bacterial Cellulose (NFBC) Produces a Promising Therapeutic Outcome for Peritoneally Metastatic Gastric Cancer in Mice Models via Intraperitoneal Direct Injection. *Nanomaterials* **2021**, *11*, 1697. [CrossRef] [PubMed]

 *nanomaterials*

Review

Construction of Functional Materials in Various Material Forms from Cellulosic Cholesteric Liquid Crystals

Kazuma Miyagi [1],* and Yoshikuni Teramoto [2],*

1   Department of Forest Resource Chemistry, Forestry and Forest Products Research Institute, Forest Research and Management Organization, 1 Matsunosato, Tsukuba 3058687, Ibaraki, Japan
2   Division of Forest and Biomaterials Science, Graduate School of Agriculture, Kyoto University, Kitashirakawa Oiwake-cho, Sakyo-ku, Kyoto 6068502, Japan
*   Correspondence: miyagik298@ffpri.affrc.go.jp (K.M.); teramoto.yoshikuni.3e@kyoto-u.ac.jp (Y.T.)

Abstract: Wide use of bio-based polymers could play a key role in facilitating a more sustainable society because such polymers are renewable and ecofriendly. Cellulose is a representative bio-based polymer and has been used in various materials. To further expand the application of cellulose, it is crucial to develop functional materials utilizing cellulosic physicochemical properties that are acknowledged but insufficiently applied. Cellulose derivatives and cellulose nanocrystals exhibit a cholesteric liquid crystal (ChLC) property based on rigidity and chirality, and this property is promising for constructing next-generation functional materials. The form of such materials is an important factor because material form is closely related with function. To date, researchers have reported cellulosic ChLC materials with a wide range of material forms—such as films, gels, mesoporous materials, and emulsions—for diverse functions. We first briefly review the fundamental aspects of cellulosic ChLCs. Then we comprehensively review research on cellulosic ChLC functional materials in terms of their material forms. Thus, this review provides insights into the creation of novel cellulosic ChLC functional materials based on material form designed toward the expanded application of cellulosics.

Keywords: cellulose derivatives; cellulose nanocrystal; cholesteric liquid crystal; functional materials; material form

Citation: Miyagi, K.; Teramoto, Y. Construction of Functional Materials in Various Material Forms from Cellulosic Cholesteric Liquid Crystals. *Nanomaterials* **2021**, *11*, 2969. https://doi.org/10.3390/nano11112969

Academic Editor: Miguel Gama

Received: 1 October 2021
Accepted: 3 November 2021
Published: 5 November 2021

Publisher's Note: MDPI stays neutral with regard to jurisdictional claims in published maps and institutional affiliations.

1. General Introduction: Need for and Overview of Cellulosic Liquid-Crystals as Functional Materials

1.1. Introduction

Cellulose is the most abundant bioderived organic polymer. Because of its excellent physical properties, including its mechanical and sorption properties, cellulose is widely used in areas such as the manufacture of paper, textiles, and structural components. In addition to conventional utilization, developing applications of unused but attractive physicochemical properties of cellulose is important to realize the wealth of a sustainable society because cellulose is reproducible and eco-friendly.

Liquid crystallinity is one of the characteristic physical properties of cellulosic molecules, although it is not fully employed in the real world. The high stimuli-responsivity and high-order structure of liquid crystals may enable cellulosic functional materials. It is also interesting that the liquid crystal property develops spontaneously. Numerous fundamental studies on cellulosic liquid crystals have been reported. The creation of functional materials that take advantage of the insights gained in those studies may contribute to expanding the areas in which cellulosics can be used.

There are several excellent reviews on the fundamental science and possible applications of cellulosic liquid-crystals. For instance, Gray [1] summarized initial research on the liquid-crystallinity of several cellulose derivatives. Lagerwall et al. [2] recently reviewed the phase diagrams of aqueous cellulose nanocrystal (CNC) dispersions and discussed ordered assembly of CNCs in suspensions as well as films. The Lagerwall group has most recently published a significantly detailed review on equilibrium and kinetically arrested liquid crystalline structure based on CNC [3]. Giese et al. [4] reviewed cellulosic liquid crystalline structures in terms of their structural templates. Moreover, a review by Nishio et al. [5] surveyed both the fundamental aspects and potential applications of liquid crystalline cellulose derivatives and nanocrystals in terms of their stimuli responses. However, to our knowledge there is no review on cellulosic liquid crystalline functional materials in the context of material forms.

The material form is an important factor involving the function and application of materials. For example, films are applied for electronic and optical devices, hydrogels are compatible with biomedical applications, and porous materials are applied for water purification. In the present review, we therefore categorize previous studies on cellulosic liquid crystalline functional materials in broad terms to solid and liquid types, and comprehensively review various material forms corresponding to the respective types.

1.2. Cellulosic Liquid-Crystals

We here briefly describe the basic fundamentals of cellulosic liquid-crystals before considering their applications as functional materials.

Cellulose has a backbone chain consisting of β-1,4-linked glucose units that impart (semi-)rigidity and chirality. These structural features endow cellulosics with the potential to form a cholesteric liquid crystal (ChLC) phase. ChLC is often modeled as being formed from stacked virtual planes, each of which is composed of molecules or particles with a uniaxial orientation. The vector indicating such orientation is called director. The director of each plane (termed as pseudonematic plane) twists periodically around the normal of the stacked pseudonematic planes, rendering a helical nature to ChLC. Figure 1a presents a schematic illustration of ChLC with polysaccharides as component molecules. Although the schematic ChLC in Figure 1a seems to be a discretely layered structure, pseudonematic planes should be continuously stacked in actual ChLC. Figure 1b is a schematic diagram focusing on the short-range structure of a ChLC to emphasize such continuously stacked pseudonematic planes.

ChLCs can be attained for rigid and chiral particles or molecules in the dispersed or solution state (at a sufficiently high concentration) termed a lyotropic ChLC, or in a molten state termed a thermotropic ChLC. The classification of lyotropic and thermotropic states applies not only to ChLCs but also to liquid-crystals in general.

ChLCs exhibit a unique optical property: selective reflection of circularly polarized light (CPL) with a wavelength equivalent to the helical pitch and with a rotation direction that is the same as that of the helix. The reflection wavelength (λ_M) and the helical pitch (P) are related by de Vries' equation, $\lambda_M = \bar{n}P$ ($\bar{n}$ is the refractive index) [6]. Accordingly, ChLCs exhibit circular dichroism (CD) as well as coloration, when P is comparable with a wavelength in the range of visible light. As stated in this equation, the cholesteric helical pitch P of ChLC samples can be calculated from their λ_M and $\bar{n}$ values, which can be acquired with a spectrophotometer and refractometer, respectively. The P value can also be estimated by microscopy; polarized optical microscope (POM) and scanning electron microscope (SEM) images of ChLC samples demonstrate the characteristic fingerprint-like texture (Figure 1c), and the distance between the adjacent dark lines corresponds to half of P. It should be noted that SEM can be used to estimate the P of solid-state ChLC samples, but not for solution-state (lyotropic) ones. Solidification of such solution samples by polymerization of their solvents described in Section 2.1 enables the measurement of P from an SEM image, although the solidification process can affect P to some extent. In accordance with the ChLC structure indicated in Figure 1a,b, P can be described as $P = 360°d/\varphi$, where d is the interplanar spacing of the stacked pseudonematic planes and φ is the twist angle in degrees. This indicates that P and thus λ_M change with d and φ. The value of d can be obtained from wide-angle X-ray diffraction analysis as shown in Figure 1d by approximating the short-range order of a ChLC with the hexagonal model [7]. The helical handedness of a ChLC is determined by CD or optical rotatory dispersion (ORD) spectroscopy. ChLC samples with a left-handed helix exhibit a positive peak in CD spectra and a positive Cotton effect in ORD spectra, whereas ChLC samples with a right-handed helix exhibit a negative peak in CD spectra and a negative Cotton effect in ORD spectra (Figure 1e). The helical handedness can also be discriminated by visual inspection of ChLC samples with circular polarizers, because left-handed and right-handed CPL (L-CPL and R-CPL, respectively) reflected from the samples can transmit through a left-handed circular polarizer (LCP) and a right-handed circular polarizer (RCP), respectively.

Expression of lyotropic ChLC requires rigid and chiral molecules or particles to have good solubility or dispersity in solvents, and thermotropic ChLC requires them to have good meltability. The native cellulose molecule has poor solubility and thermal meltability because of its strong inter- and intramolecular hydrogen bonding. Instead, cellulose derivatives and nanocrystals have typically been adopted to achieve a cellulosic ChLC phase.

The CNC is the structural unit of a cellulose crystal and can be produced by acid hydrolysis of cellulose microfibrils [8,9]. Cellulose microfibrils in wood are generally considered to be composed of bundles of 6×6 cellulose molecular chains, although there are recent reports on cellulose microfibrils comprising 18 molecular chains [10]. Such cellulose bundles exhibit alternating amorphous and crystalline regions along the molecular chain direction. Acid hydrolysis of cellulose bundles preferentially degrades the amorphous regions. After degrading most of the amorphous regions, the degradation is substantially decelerated, resulting in residual highly crystalline particles which are the CNCs. According to this characteristic, the degree of polymerization of cellulose chains in the crystalline regions is called the level-off degree of polymerization (the degradation "turns off" at this degree of polymerization) [11]. CNCs prepared by hydrolysis with sulfuric acid are an especially active area of research. CNCs can express lyotropic ChLC in aqueous suspensions based on CNCs' good water-dispersity, which is attributable to the introduced surface sulfate ester groups [12,13]; whereas CNCs generally do not express thermotropic ChLCs because of CNCs' lesser meltability. Since the first report of liquid-crystallinity in CNCs [14], fundamental research on CNC-based ChLCs have been made [15–20]. CNC-based ChLCs typically form a left-handed helix, and to our knowledge there has been no report on formation of a right-handed helix from CNCs to date.

Cellulose derivatives can be obtained by substituting the hydroxy groups in cellulose, typically by etherification or esterification [21]. Because substituents suppress the strong intra- and intermolecular hydrogen bonding of cellulose, cellulose derivatives often exhibit good solubility and meltability. Such properties facilitate expression of lyotropic and thermotropic ChLCs. The first reports of formation of ChLCs by cellulose derivatives were for an aqueous hydroxypropyl cellulose (HPC) solution system [22]. Since then, many important fundamental studies on cellulose derivative-based ChLCs have been presented— such as the effects of concentration, temperature, solvent, side chain structure, and degree of substitution on the selective reflection properties [23–29].

In contrast to CNCs, left- and right-handed cholesteric helices have been observed for cellulose derivative-based ChLCs. Most recently, Nishio et al. [30,31] proposed that the effects of the substituents and temperature on the helical handedness of cellulosic ChLCs can be systematically interpreted in terms of the balance between the dispersion interaction and steric repulsion, based on their experimental data and the theoretical work of Osipov [30,31].

There is a difference between the lyotropic ChLCs of cellulose derivatives and CNCs. For the former, the building blocks of the ChLC structure are molecules of cellulose derivatives, and the component molecules completely dissolve in solvents. Conversely, for the latter, CNC "particles" that disperse in solvents (rather than dissolve) provide the ChLC structure. Accordingly, ChLC systems comprising cellulose derivatives and CNCs are sometimes termed molecular ChLCs and colloidal ChLCs, respectively. The structural difference between cellulose derivatives and CNC leads to different characteristics of their ChLC samples. The representative examples include instances where the cholesteric helical pitch of cellulose derivative-based ChLCs is shorter than those of CNC's with corresponding concentration because of the smaller diameter of cellulose derivatives (~1 nm) relative to CNC (5–20 nm [8,32]); and where the critical concentration to express lyotropic ChLC for CNC is lower than that of cellulose derivatives because of the very rigid crystal backbone of CNC.

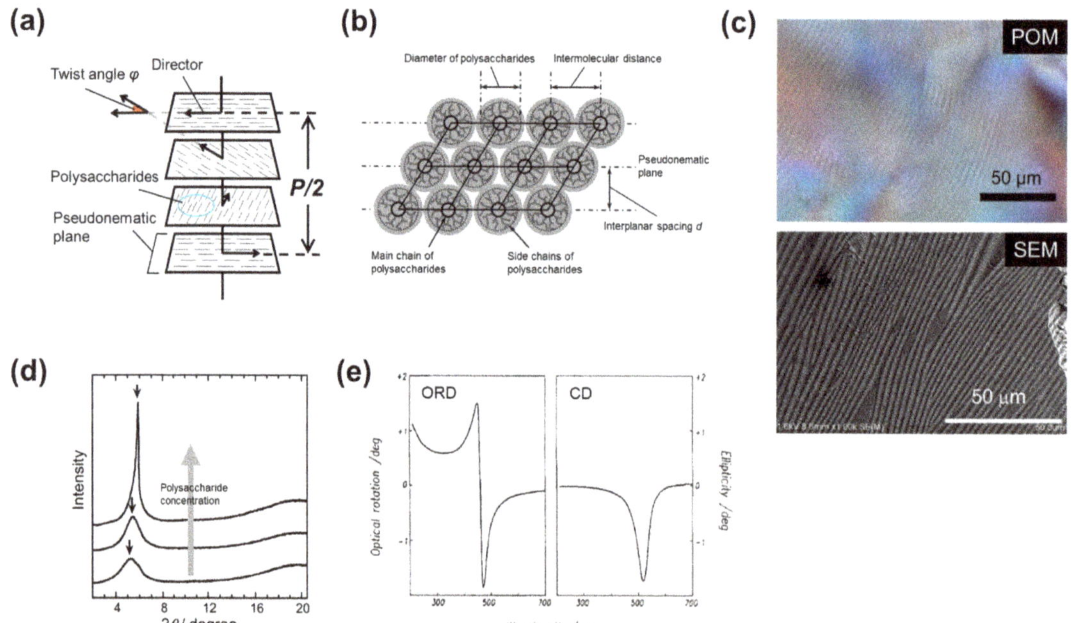

Figure 1. (**a**) Schematic that illustrates the helical pitch P, director, twist angle φ for a cholesteric liquid crystalline (ChLC) phase composed of polysaccharides. (**b**) Schematic illustrating the pseudonematic planes and interplanar spacing d for a short-range structure of a polysaccharide-based ChLC phase (reprinted with permission from [7] Copyright (2021) American Chemical Society). (**c**) Representative polarized optical microscope (POM) image of solution-state polysaccharide-based ChLC sample and scanning electron microscope (SEM) image of solid-state sample (reprinted with permission from [33] Copyright (2021) American Chemical Society). (**d**) Representative wide-angle X-ray diffraction profiles of solution-state polysaccharide-based ChLC samples with different polysaccharide concentrations (reprinted with permission from [7] Copyright (2021) American Chemical Society). (**e**) Representative circular dichroism (CD) and optical rotary dispersion (ORD) spectra of a right-handed ChLC sample (Reprinted with permission from [34] Copyright (2021) American Chemical Society).

Various insights accumulated by previous fundamental studies on cellulosic ChLCs have led to increased research activity dedicated to processing ChLCs into material forms intended for practical use. The present article reviews information on functional materials with diverse material forms produced from cellulose derivative-based and CNC-based ChLCs. The following review is divided into sections based on material forms, and we concurrently review both cellulose derivative-based and CNC-based functional materials. Please note that the difference in characteristics between the aforementioned ChLCs leads to different experimental conditions and procedures to construct the functional materials even in the same material forms.

2. Cellulosic ChLC-Based Solid Functional Materials

2.1. Cellulosic ChLC Films

Cellulose derivatives and CNC are useful components for producing films incorporating a ChLC structure. Representative methods for preparing cellulose derivative-based ChLC films are solution-casting [35–37] and in situ polymerization [38–43] of a lyotropic ChLC solution of cellulose derivatives. In solution-casting methods, typically the cellulosic ChLC solutions are first spread into a film-like shape and sandwiched (i.e., packed) between two glass plates, followed by slow evaporation of the solvents. Although this method is operationally simple, it insufficiently immobilizes ChLC structure with a long helical pitch P corresponding to a red wavelength because the cholesteric helical pitch decreases in accordance with increasing cellulosic concentration of the solutions over the course of drying. However, it is straightforward to immobilize a ChLC structure with a short P in considerably concentrated cellulosic solutions because of the extremely low mobility of the cellulosic chains. Regarding the in situ polymerization method, lyotropic ChLC solutions are prepared by dissolving cellulose derivatives in polymerizable solvents (e.g., acrylate and acrylamide monomers) and polymerization-initiator, and then the solution samples—molded between the glass plates—are subjected to heating or ultraviolet (UV) irradiation to initiate polymerization of the monomeric solvents. The ChLC structure formed in the solution state is retained even after the monomeric solvents are converted to polymeric matrices (e.g., polyacrylate and polyacrylamide), and these matrices help to further immobilize the ChLC structure. Thus, in situ polymerization methods can immobilize ChLC structures that have not only a short P but also a long P, unlike solution-casting methods. This renders in situ polymerization a general method for preparing cellulose derivative-based ChLC films. However, the ability to immobilize the ChLC structure strongly depends on compatibility between cellulose derivatives and the resultant synthetic polymers, because interaction of mesogenic cellulosic chains may be distorted when cellulose derivatives completely mix with the synthetic polymers, as described in Section 2.1.1.

For CNC-based ChLC films as well, solution (dispersion)-casting [2,12,44] and in situ polymerization [4,5] are representative preparation methods. In contrast to cellulose derivative-based ChLC films, not only in situ polymerization but also casting is frequently adopted for CNC-based films. In colloidal ChLCs formed by CNCs, the length scale of the interplanar spacing of the pseudonematic planes is longer than that of molecular ChLCs formed by cellulose derivatives. As a result, the cholesteric helical pitch of the CNC ChLC dispersion remains in the visible-wavelength region, even at high concentration. The ChLC structure with concentrated dispersion is then immobilized in the obtained film after complete evaporation of the solvent, because of the low mobility of CNC rods under high concentration. Whereas cellulose derivative-based ChLC solutions can be prepared by simply dissolving dried cellulose derivatives in solvents, completely dispersing dried CNC in solvents is difficult. Therefore, CNC-based lyotropic ChLCs are prepared by concentrating dilute CNC dispersions obtained after acid hydrolysis (and purification) of cellulose. The typical concentration methods are solvent evaporation [4] and dialysis [14,45].

2.1.1. Mechanochromic Films

Materials that exhibit a color change in response to mechanical stimuli—i.e., mechanochromic materials—have applications as e.g., mechanical damage sensors and energy-saving display devices. A representative approach to developing mechanochromic materials is using structural color [46]. Structurally colored materials present color that is typically ascribed to Bragg reflections from the materials' internal periodic layered structures, and this color can change when a given mechanical stimuli varies the periodicity [47–50]. ChLCs also exhibit structural color that can change in accordance with the helical periodicity, and thus ChLCs can be key structures for constructing structurally colored mechanochromic materials. In addition, ChLCs also exhibit CD and thus optical chirality. This unique property of ChLCs may bring new concepts to the field of mechanochromic materials.

Müller and Zentel [51] prepared structurally colored mechanochromic films from cellulose derivative-based ChLC. They prepared lyotropic ChLC solutions by dissolving 44 wt% cellulose 3-chlorophenylcarbamate (Figure 2a) in acrylate monomer with crosslinker, and then subjecting the solutions to in situ thermal polymerization. The researchers compressed the ChLC film between two glass slides and evaluated the mechanochromic properties of the film by UV–visible (Vis) spectroscopy. The selective reflection wavelength of the ChLC film is blue-shifted by compression (Figure 2b). Furthermore, thermal treatment of the film recovered the original reflection wavelength.

Liang et al. [52] reported large-area photonic films from HPC by the roll-to-roll method. They produced the photonic films by laminating HPC in aqueous ChLC but did not conduct immobilization through solution-casting or in situ polymerization. The roll-to-roll fabrication consisted of coating the lyotropic ChLC on a polyethylene terephthalate substrate, glue-deposition to seal the edge of the film, UV-curing the glue, and laminating the film (Figure 2c). The laminated ChLC films exhibited a decrease in the reflection wavelength in accordance with increasing applied compression stress. The reflection spectra broadened by compression, probably because of misalignment of the ChLC domain induced by lateral shear flow. The researchers presented an excellent touch-sensing function of the laminated ChLC films through visualizing footprints by color change (Figure 2d).

Boott et al. [53] developed mechanochromic elastomeric films by compounding CNC-based ChLC films with synthetic polymers. They cast 4 wt% CNC aqueous suspensions to conduct EISA to obtain CNC ChLC films. The researchers then soaked the films with a dimethylsulfoxide (DMSO) solution of ethyl acrylate, 2-hydroxyethyl acrylate, and initiator to infiltrate these reagents into the films. They heated the vial to initiate polymerization of the acrylate monomers adsorbed in the films, resulting in CNC-based ChLC elastomeric films.

UV–Vis spectra of the stretched ChLC elastomeric film indicate that the film's color blue-shifted in accordance with increasing tensile strain (Figure 2e). This shift indicates a decrease in the cholesteric helical pitch of the film by stretching. CD spectra also supported this result and exhibited a positive CD signal—demonstrating preservation of CNC-based left-handed ChLC in the stretched film (Figure 2f). Relaxation recovered the original color of the film. To evaluate the effect of the tensile stress on the internal ChLC structure, Boott et al. [53] conducted two-dimensional X-ray diffraction analyses for the relaxed and stretched sample. They observed a diffraction peak corresponding to unwinding of the ChLC structure into a pseudonematic structure when they stretched the sample. Accordingly, the researchers proposed a structural model of CNC assembly in a stretched film whereby some of the CNC rods were almost parallel to the stretch direction before stretching, and these CNC rods completely aligned with the stretched direction upon application of stress, giving rise to a small nematic region (Figure 2g).

Whereas most studies on cellulosic ChLC-based mechanochromic materials investigated force-induced color changes, Miyagi and Teramoto [54] focused on an effect of mechanical stimuli on the CD of cellulosic ChLC films. They prepared ethylcellulose/poly(acrylic acid) (EC/PAA) ChLC films by dissolving 47 wt% EC in AA solvent with photoinitiator followed by in situ photopolymerization of AA.

Miyagi and Teramoto [54] subjected 47 wt% EC/PAA ChLC films to compression at 130 °C, which is higher than the glass transition temperature (T_g) of the films, and measured the CD spectra of the compressed films. The original (i.e., not compressed) EC/PAA films exhibited a positive CD signal, indicative of selective reflection of L-CPL; whereas the compressed films exhibited a negative CD signal corresponding to selective reflection of R-CPL (Figure 2h). In addition, the researchers visually inspected the EC/PAA films with circular polarizers before and after the compression. The reflected light from the original films transmitted through LCP, whereas the reflected light from the compressed films passed through RCP; which is consistent with the spectroscopic measurements (Figure 2i). These CD and circular polarization results are indicative of stress-induced circular dichroic inversion (SICDI) of the ChLC films. The researchers hypothesized that the mechanism of this circular dichroic inversion is as follows: the component polymers of the ChLC films were partially linearly oriented, resulting in birefringence of the transmitted CPL that was to undergo handedness inversion. They derived a mathematical model, taking this hypothesis into account, to calculate theoretical CD spectra of the compressed EC/PAA films by modifying the equation reported by Ritcey et al. [55]. The obtained theoretical CD spectra of the compressed films exhibited good agreement with corresponding experimental spectra (Figure 2j), indicating the validity of the researchers' hypothesis for the mechanism of SICDI [56].

In prior work EC/PAA films expressed SICDI only above 130 °C because of the T_g of the films. Thus Miyagi and Teramoto [57–59] also expanded the conditions where such films can manifest SICDI. Accordingly, they prepared propionylated HPC (PHPC; Figure 2k), which has a low T_g (ca. −20 °C) and high ChLC-formability in a wide range of monomeric solvents. In this system, the ability to immobilize ChLC structures into films by in situ polymerization depended on the species of the in situ-generated polymers. Fukawa et al. [60] reported a similar phenomenon. Spin-lattice relaxation analyses of PHPC/synthetic polymer ChLC films by solid-state nuclear magnetic resonance spectroscopy indicated that the immobilization efficiency was high for the films where PHPC and synthetic polymers were not completely compatible [57,58]. This can be attributed to the way that the complete mixing of cellulose derivatives and synthetic polymers disturbs the interaction of mesogenic cellulosic chains. PHPC/synthetic polymer films with an immobilized ChLC structure exhibited SICDI by compression at room temperature (Figure 2l).

2.1.2. Biomineralized Films

Living organisms build structurally strong tissues in their bodies by compounding organics and minerals—so-called organic–inorganic hybrids, such as vertebrate bones and teeth, fish scales, crustacean exoskeletons, and shells. Representative minerals are e.g., hydroxyapatite, calcium carbonate, and magnetite. Inspired by these natural structures, functional material designs based on biomineralization are an active area of research [61,62]. There has been research on biomineralization of cellulosic ChLC films, as discussed next.

Ogiwara et al. [63] conducted calcium phosphate biomineralization using a cellulose derivative-based ChLC as an organic scaffold. This concept likely mimics the mineralized supramolecular helical structure that is seen in the crustacean exoskeleton. Although they investigated EC and HPC-based ChLC films, we focus on only the EC system. The researchers prepared EC/PAA films by in situ photopolymerization of 40–57 wt% EC/AA ChLC solutions and soaked the obtained films in mineral solutions containing $CaCl_2$, $(NH_4)_2HPO_4$, and NaCl at pH 9.0. They then rinsed the mineralized films with water and vacuum-dried the films. In this material design, PAA is a crucial component that has the two roles of immobilizing the cellulosic ChLC structure and absorbing the ions, thus facilitating mineralization on the ChLC scaffold.

Energy-dispersive X-ray analyses of the fracture surface of mineralized 55 wt% EC/PAA films (*m*-EC(55)/PAA) indicate that the elements Ca and P were uniformly distributed inside the films by mineralization (Figure 3a). Wide-angle X-ray diffraction profiles of unmodified 55 wt% EC/PAA films (EC(55)/PAA; Figure 3b) showed a diffraction peak that is attributable to the stacked pseudonematic planes of ChLC at $\theta = 9.6°$ and a halo corresponding to an amorphous region at $\theta \approx 19°$. Both the unmodified and as-mineralized EC/PAA films exhibited no clear crystalline diffraction pattern, whereas the mineralized films subjected to posttreatment in aqueous NaOH (p-*m*-EC(55)/PAA) exhibited diffraction peaks from crystalline hydroxyapatite (Figure 3b). The NaOH posttreatment results are probably attributable to alkali hydrolysis of amorphous calcium phosphate or less-ordered precursors of hydroxyapatite.

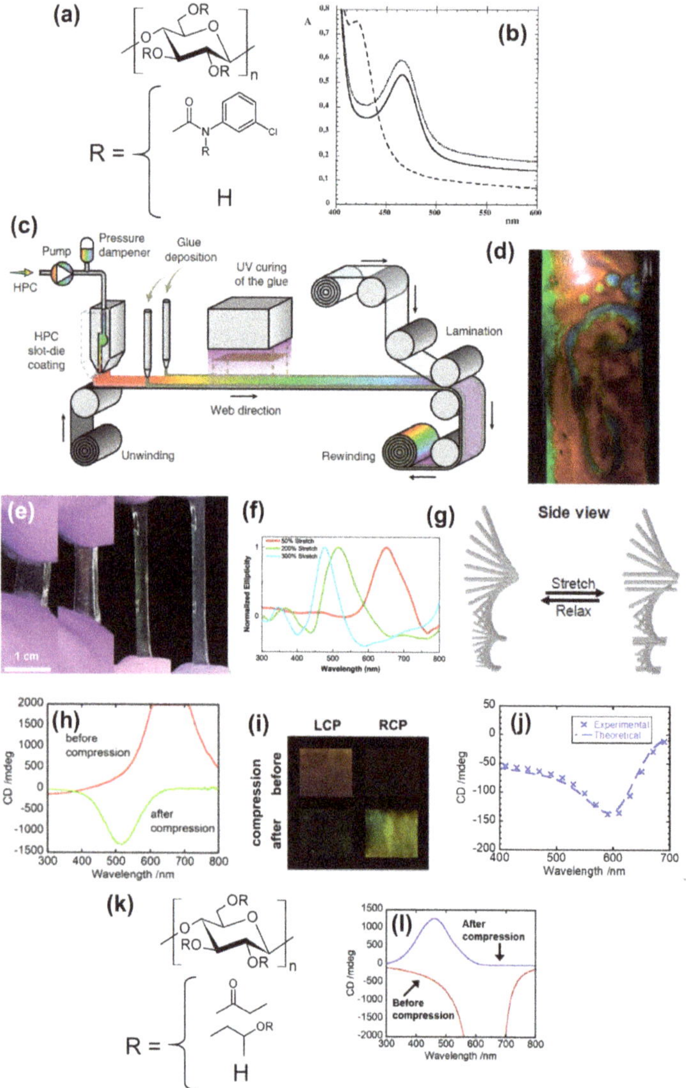

Figure 2. (**a**) Chemical structure of cellulose 3-chlorophenylcarbamate and (**b**) ultraviolet (UV)–visible spectra of a cholesteric liquid crystalline (ChLC) film before compression (solid line), directly after compression (dashed line), and after heating at 50 °C (dotted line) (reproduced from [51] by permission of Wiley). (**c**) Schematic of roll-to-roll preparation of hydroxypropyl cellulose (HPC) ChLC films and (**d**) footprint exhibited by the films (adapted from [52] by permission of Springer Nature). (**e**) Visual appearances and (**f**) circular dichroism (CD) spectra of a stretched cellulose nanocrystal ChLC elastomeric film. (**g**) Proposed model of a change in the ChLC structure in the film under tensile ((**e–g**) were reproduced from [53] by permission of Wiley). (**h**) CD spectra of ethylcellulose/poly(acrylic acid) (EC/PAA) ChLC films before and after compression at 130 °C and (**i**) visual appearance through circular polarizers (LCP: left-handed circular polarizer; RCP: right-handed circular polarizer) (reproduced from [54] by permission of the Royal Society of Chemistry). (**j**) Experimental and theoretical CD spectra of compressed EC/PAA films (adapted with permission from [56], copyright (2021) American Chemical Society). (**k**) Chemical structure of propionylated HPC (PHPC) and (**l**) CD spectra of PHPC/poly(methyl methacrylate) ChLC films before and after compression at 30 °C (reproduced from [58] by permission of Elsevier).

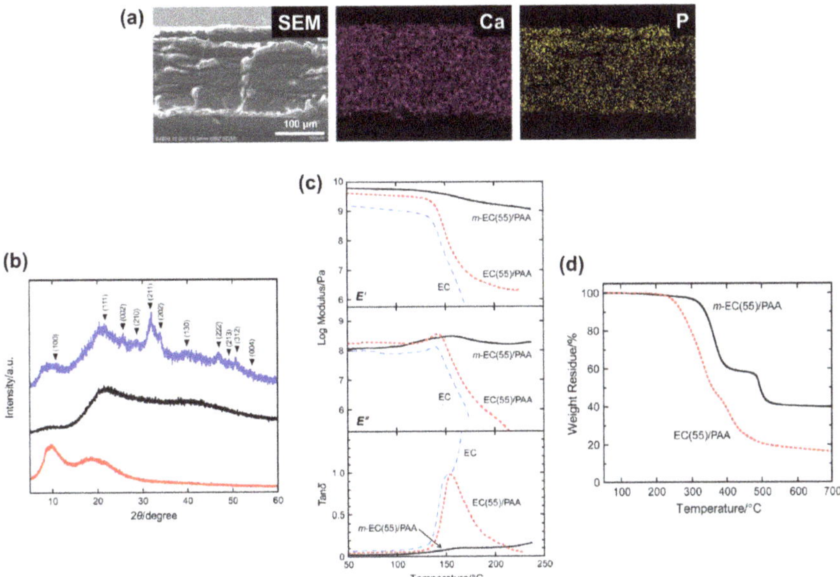

Figure 3. (**a**) Scanning electron microscope (SEM) image and energy-dispersive X-ray mappings of Ca and P for the fracture surface of an *m*-EC(55)/PAA film. (**b**) Wide-angle X-ray diffraction intensity profiles of EC(55)/PAA (bottom), *m*-EC(55)/PAA (center), and p-*m*-EC(55)/PAA film (top). (**c**) Temperature dependence of the storage modulus *E′*, loss modulus *E″*, and loss factor tan *δ* for EC (dashed line), EC(55)/PAA (dotted line), and *m*-EC(55)/PAA film (solid line). (**d**) Thermogravimetric analysis profiles for EC(55)/PAA and *m*-EC(55)/PAA films ((**a**–**d**) were reprinted with permission from [63], Copyright (2021) American Chemical Society).

In dynamic mechanical analyses of the film samples, the storage and loss modulus of EC(55)/PAA films substantially diminished from 140 °C, whereas those of *m*-EC(55)/PAA films retained relatively high values even above 200 °C (Figure 3c). Thermogravimetric analyses of these samples indicate that the temperature at which polymer degradation began and the residual weight at 700 °C for *m*-EC(55)/PAA films were higher than those of EC(55)/PAA films (Figure 3d). These dynamic mechanical analysis and thermogravimetric analysis results reveal that biomineralization enhanced the thermomechanical properties and thermal stability of EC/PAA ChLC films.

Katsumura et al. [64] and Nakao et al. [65] reported calcium carbonate mineralization of EC/PAA, CNC/poly(2-hydroxyethyl methacrylate), and CNC/poly(HEMA-*co*-AA) ChLC films. All of these systems exhibited improved thermomechanical properties and thermal stability, as well as calcium phosphonate mineralization system.

2.1.3. Optical Filters

Although ChLC can act as an optical filter to separate the CPL of particular wavelengths and handedness, this reflection-selectivity is sometimes too specific. However, by combining ChLC with a nematic liquid crystalline phase or substances that affect the cholesteric helical structure, the reflection-selectivity can be expanded. This enables fabrication of optical filters with a filtering function tailored to assumed applications.

De la Cruz et al. [66] produced optical filters that reflect both L-CPL and R-CPL with a certain wavelength, by combining ChLC films and nematic liquid crystalline films, both prepared from CNCs. These filters have a reflection-selectivity for wavelength and no selectivity for handedness. The researchers designed the optical filters by a sandwich structure with top and bottom left-handed ChLC CNC layers, and a middle nematic CNC layer as a wave-plate (Figure 4a). In this system, the first ChLC layer selectively reflected

L-CPL, and the birefringence inverted the transmitted R-CPL in the middle nematic layer. The L-CPL thus generated was reflected from the second ChLC layer, inverted back into R-CPL in the middle layer, and transmitted through the first ChLC layer. This mechanism is analogous to the SICDI by the compressed EC/PAA ChLC films described in Section 2.1.1.

De la Cruz et al. [66] prepared CNC-based ChLC films by curing mixtures of a CNC aqueous dispersion, 1,2-bis(trimethoxysilyl)ethane (organosilica precursor), and low-molecular-weight polyethylene glycol (plasticizer) in ambient conditions. Curing was in the presence of a static magnetic field normal to the film surface for uniform orientation of the helical axis: i.e., perpendicular to the film surface. The helical pitch, and thus a selective reflection wavelength, of the films can be tuned by varying the feed ratio of the CNC aqueous dispersion, organosilica, and polyethylene glycol. The researchers prepared CNC-based nematic films by linear shear deposition of a 3.5–5.0 wt% CNC aqueous dispersion on the ChLC films, and subsequent drying at 30 °C to 35 °C. They prepared sandwich films by attaching the ChLC film with the nematic layer to the pristine ChLC film with an adhesive.

UV–Vis reflection spectroscopy of the single ChLC CNC layer exhibited no reflection of R-CPL because of the selective reflection of L-CPL. In addition, the reflectance of linearly polarized (LP) light was half that of L-CPL because LP light is the vector sum of R-CPL and L-CPL (Figure 4b). In contrast, the sandwich film exhibited almost the same reflectivity for LP light, L-CPL, and R-CPL; indicating the handedness-independent reflection of CPL (Figure 4b). This interpretation is supported by CD spectra that exhibited no peak for the sandwich film because of equal reflection of both L-CPL and R-CPL, and transmission spectra where the minimum transmittance of the sandwich film was half that of the single-ChLC film. Such handedness-independent optical filters can achieve full reflection of both L-CPL and R-CPL, enabling twice the efficiency of the resulting photoenergy in comparison with single-ChLC films.

Fernandes et al. [67] fabricated CNC ChLC films that also reflect both R-CPL and L-CPL, by incorporating a small-molecule nematic liquid-crystal in the films. Although the concept of this material to accomplish handedness-independent reflection is the same as in de la Cruz et al. [66], their material designs are different. Whereas de la Cruz et al. [66] constructed the optical filters from completely separated ChLC films and nematic liquid crystalline films, Fernandes et al. [67] integrated the ChLC films and nematic liquid-crystal as one material system.

Fernandes et al. [67] prepared ChLC films by casting a 1.9 wt% CNC aqueous dispersion on a Petri dish and drying the dispersion at 20 °C for 4 w to allow EISA. They prepared sample cells by placing the ChLC films between two glass slides and filled the sample cells with the nematic liquid-crystal 4-pentyl-4′-cyanobiphenyl (5CB) to obtain CNC–5CB composite liquid crystalline films.

Fernandes et al. [67] observed the pristine CNC ChLC films and the CNC–5CB composite films with circular polarizers. Whereas the CNC films exhibited the reflection color in the LCP image but no color in the RCP image, the composite films exhibited the reflection color in both the LCP and RCP images (Figure 4c). These differing reflection results indicate that the composite films reflected both L-CPL and R-CPL. Analogously to de la Cruz et al.'s [66] strategy, the birefringence in the nematic region inverted the R-CPL transmitted through the outer ChLC region, reflected this inverted transmission from the inner ChLC region, reinverted this reflection as R-CPL in the nematic region, and transmitted the reinversion through the outer ChLC region.

Because 5CB is a small-molecule liquid crystal, it has high stimuli-sensitivity and a fast response. Taking advantage of this characteristic, the researchers can modulate the reflection properties of CNC–5CB composites by external stimuli—such as the temperature and electric field. The composite films exhibited handedness-independent reflection of CPL at 30 °C, whereas they exhibited selective reflection of L-CPL at 34.5 °C (Figure 4d). This is attributable to 5CB not exhibiting birefringence because of the nematic–isotropic phase transition at ~35 °C. Applying an electric field parallel to the cholesteric helical axis eliminated the reflection of R-CPL (Figure 4e), because no birefringence occurs when the orientation direction of an anisotropic material is parallel to the light direction. The CNC–5CB composite is promising for applications as an optical filter that can dynamically switch their filtering mode.

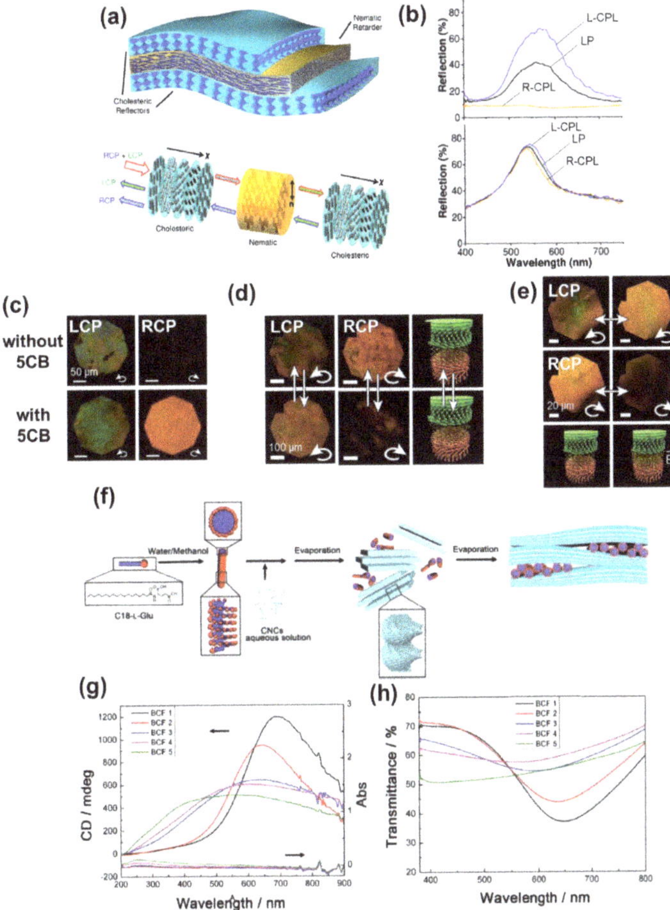

Figure 4. (**a**) Schematic describing the sandwich film composed of cellulose nanocrystal (CNC)-based cholesteric liquid crystal (ChLC) outer layers and a nematic inner layer, and the mechanism of the handedness-independent circularly polarized light (CPL) reflection. LCP, left-handed circular polarizer; RCP, right-handed circular polarizer. (**b**) Reflection spectra of a single CNC ChLC film (top) and a sandwich film (bottom) with linearly polarized (LP) light, right-handed CPL (R-CPL), and left-handed CPL (L-CPL) incident light ((**a**,**b**) were reprinted with permission from [66], copyright (2021) American Chemical Society). (**c**) Reflection images of a simple CNC ChLC film and a CNC–4-pentyl-4′-cyanobiphenyl (5CB) composite through LCP and RCP. (**d**) Reflection image of a CNC–5CB composite through LCP and RCP at 30 °C (top) and 34.5 °C (bottom). (**e**) Reflection image of a CNC–5CB composite through LCP and RCP; before (left) and after (right) applying an electric field parallel to the cholesteric helical axis ((**c**–**e**) were reproduced from [67] by permission of Wiley). (**f**) Schematic presenting formation of the ChLC structure with a distributed helical pitch in the presence of *N*-stearoyl-L-glutamic acid (C18–L-Glu micelles). (**g**) Circular dichroism spectra and (**h**) transmission spectra of broadband cellulose films (BCFs) with various portions of C18–L-Glu ((**f**,**g**) were reproduced from [68] by permission of Wiley).

Cao et al. [68] developed optical filters that reflect L-CPL with a wide range of wavelengths in the visible region, by combining ChLC films with a surfactant. These films have reflection-selectivity for the handedness and no selectivity for the wavelength. The concept of the material design is that nanometer-scale micelles formed from the surfactant disturb the self-assembly of CNC to give a nonuniform cholesteric helical pitch, resulting in broadband reflection of L-CPL (Figure 4f).

Cao et al. [68] obtained CNC-based ChLC films compounded with micelles by adding a controlled quantity of micelle solution composed of a water/methanol mixture and anionic surfactant *N*-stearoyl-L-glutamic acid (C18–L-Glu) into a 4 wt% CNC aqueous dispersion at neutral pH, followed by EISA at room temperature for 3 d. The researchers termed the prepared films as broadband cellulose films (BCFs).

Figure 4g shows CD spectra of BCFs with various proportions of C18–L-Glu (BCF1–BCF5; the C18–L-Glu content increases from BCF1–BCF5). These spectra indicate that the reflection peak of the films broadened in accordance with increasing C18–L-Glu concentration. The sample with the highest content of C18–L-Glu in the researchers' experiments (BCF5) exhibited a broad reflection peak covering the visible region. CPL transmission spectra verified the prevention of the transmission of L-CPL over a wide range of wavelengths by ChLC films incorporating C18–L-Glu (Figure 4h). Such ChLC films that exhibit broadband selective reflection of CPL can be applied to optical filters that separate R-CPL and L-CPL regardless of wavelength.

2.1.4. Conducting Films

Lizundia et al. [69] fabricated conducting materials by depositing a conducting polymer, polypyrrole (PPy), on CNC ChLC films. They prepared the ChLC films by casting 4 wt% CNC aqueous dispersions on Petri dishes, and then air-drying the dispersions for 96 h to induce EISA. To prepare CNC ChLC films with various functional groups on the surfaces, the researchers conducted surface modification of the films by (2,2,6,6-tetramethylpiperidin-1-yl)oxyl–oxidation, acetylation, alkaline desulfation, and cationization. They obtained CNC/PPy composite films by in situ oxidative polymerization of pyrrole on a series of CNC films as substrates (Figure 5a). The researchers soaked the CNC films in distilled water; then added pyrrole, aqueous $FeCl_3$ as an oxidant, and HCl as a dopant, before finally conducting the reaction at 0 °C for 5 h.

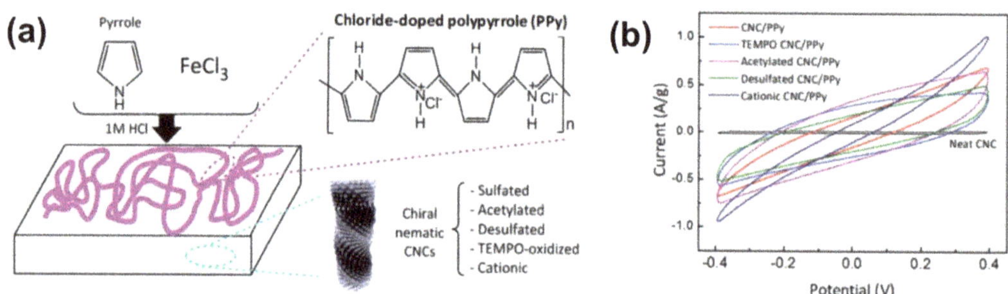

Figure 5. (**a**) Schematic for preparing cellulose nanocrystal (CNC)/polypyrrole (PPy) composite films. (**b**) Cyclic voltammograms of CNC/PPy composites, where the CNCs were subjected to various surface modifications ((**a**,**b**) were reproduced from [69] by permission of Wiley). TEMPO, (2,2,6,6-tetramethylpiperidin-1-yl)oxyl.

Lizundia et al. [69] performed cyclic voltammetry using CNC/PPy composite films as electrodes to characterize the conductivity of the films (Figure 5b). Whereas the neat CNC film without PPy exhibited zero current over the range of applied voltage, CNC/PPy composite films exhibited rectangular cyclic voltammetry curves. The specific capacitance calculated from the voltammogram of the acetylated CNC/PPy film was 4× that of the unmodified CNC/PPy film. These cyclic voltammetry results indicate that the capacitance can be improved by surface modification of composite films. The researchers considered this improvement to be related to the hydrogen bonding between the carbonyl and hydroxy groups on the CNC film, and the amine groups of the PPy.

2.2. Cellulosic ChLC Gels

A polymeric gel is composed of a polymer network constructed by chemical or physical crosslinking of the polymer chains, and swells by uptaking solvents into the network. Although Section 2.1 introduced some film materials—such as the mechanochromic films created by Müller and Zentel [51] and the biomineralized films reported by Ogiwara et al. [63] that had a polymer network structure, these materials express their functions in a dried state. We consider materials (incorporating cellulosic ChLC) that express their functions through swelling with a solvent as cellulosic ChLC functional gels.

Because cellulosic ChLC can be formed in an aqueous solution or a dispersion using HPC or CNC, these cellulosics enable production of ChLC hydrogels. This hydrogel characteristic facilitates fabrication of ecofriendly, biocompatible, cellulosic ChLC-based functional gels.

Researchers prepare cellulose derivative-based ChLC gels by in situ polymerization of lyotropic ChLC solutions of cellulose derivatives in monomeric solvents with a crosslinker [70]. Another method is formation of lyotropic ChLC solutions of cellulose derivatives with crosslinkable side groups followed by crosslinking [51]. A representative preparation of CNC-based ChLC gels is as follows: first mix a relatively dilute non-ChLC dispersion with a crosslinkable monomer, and then subject the mixture to EISA and in situ polymerization [71,72].

2.2.1. Salt-Responsive Color Gels

Chiba et al. [70] reported cellulose derivative-based ChLC hydrogels that exhibit a salt-responsive color change. They prepared lyotropic ChLC by dissolving 60–72 wt% HPC in a solvent mixture composed of di(ethylene glycol) monomethyl ether methacrylate (DEGMEM), methanol, water, and tetra(ethylene glycol) diacrylate. The researchers added glutaraldehyde and a small quantity of hydrochloric acid into the lyotropic ChLC, followed by in situ photopolymerization and crosslinking to obtain the ChLC gels. In this gel system, they immobilized the ChLC structure with an interpenetrating polymer network consisting of HPC crosslinked with glutaraldehyde, and poly(DEGMEM) (PDEGMEM) crosslinked with tetra(ethylene glycol) diacrylate.

The 65 wt% HPC/PDEGMEM composites were orange in the original dried state, whereas the color redshifted by swelling in aqueous LiSCN and subsequent drying (Figure 6a). Chiba et al. [70] restored the original color by washing the gels with distilled water and drying. However, after immersing these gels in aqueous KNO_3 the reflected color of the gels blueshifted. The researchers explained this salt-dependent color variation in terms of the chaotropic effect of the salt ions. Because chaotropic ions act as water-structure breakers (and antichaotropic ions act as water-structure formers), the salts likely affected the water-solubility of HPC, inducing an increase in the cholesteric helical pitch in the context of chaotropic ions and a decrease in the cholesteric helical pitch in the context of antichaotropic ions. Ions with a stronger chaotropicity corresponded to a larger change in the helical pitch.

In addition, Chiba et al. [70] achieved color control of the gels with an electric field (Figure 6b). After applying an electric field to HPC/PDEGMEM gels salted with LiSCN, negative thiocyanate ions migrated to the positive side of the gels, and the coloration of the positive side redshifted because of the chaotropicity of thiocyanate.

2.2.2. Humidity-Responsive Color Gels

Müller and Zentel [51] produced cellulosic molecular ChLC hydrogels that changed color in accordance with the water content. They introduced crosslinkable acryl groups into HPC by an esterification reaction between acryloyl chloride and the hydroxy groups of HPC (Figure 7a). The researchers set a relatively low degree of substitution (0.22) to maintain the water solubility. They dissolved the modified HPC in water at 66 wt% to obtain lyotropic aqueous ChLC, and then subjected the solutions to photocrosslinking to prepare ChLC hydrogels.

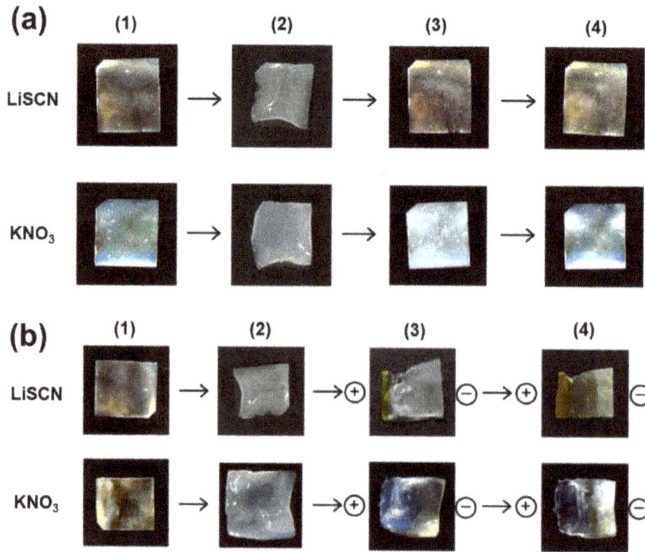

Figure 6. (**a**) Visual appearance of hydroxypropyl cellulose (HPC)/poly(di(ethylene glycol) monomethyl ether methacrylate) (PDEGMEM) gels (1) in the original state, (2) swollen in salt solutions, (3) dried after salting-in, and (4) washed with water followed by drying. The researchers used LiSCN and KNO_3 as salts. (**b**) Visual appearance of HPC/PDEGMEM gels (1) in original state, (2) swollen in salt solutions, (3) after applying the electric field, and (4) dried after turning off the electric field ((**a,b**) are reprinted with permission from [70], copyright (2021) American Chemical Society).

The obtained HPC-based ChLC hydrogel originally exhibited a red color, whereas the color blueshifted after air-drying for 3 d and shifted to the UV region after 3 w (Figure 7b). These color shifts are likely because shrinking the hydrogel by air-drying shrank the cholesteric helix incorporated in the hydrogel. The coloration of the gel returned to the red region after swelling with water, and Müller and Zentel [51] cycled the color change between the red to UV regions.

Wu et al. [71] fabricated composites (of a CNC-based ChLC gel and a polyamide film) that exhibit a humidity-responsive color change and three-dimensional deformation. They produced the composite by sandwiching a uniaxially oriented polyamide film between two CNC ChLC gel layers.

Wu et al. [71] mixed a 1.5 wt% CNC aqueous dispersion with 5 wt% aqueous polyethylene glycol diacrylate (PEGDA) to reach several weight ratios of CNC and PEGDA. They used PEGDA to construct a polymer-network structure through polymerization and to impart flexibility to the resultant polymerized product. The researchers added the CNC/PEGDA mixture to one side of a uniaxially oriented polyamide-6 (PA-6) film, and then dried the product. They subjected another side of the PA-6 film to the same process and exposed the sample to UV-irradiation for photopolymerization, resulting in a CNC ChLC gel/PA-6 film composite (Figure 7c).

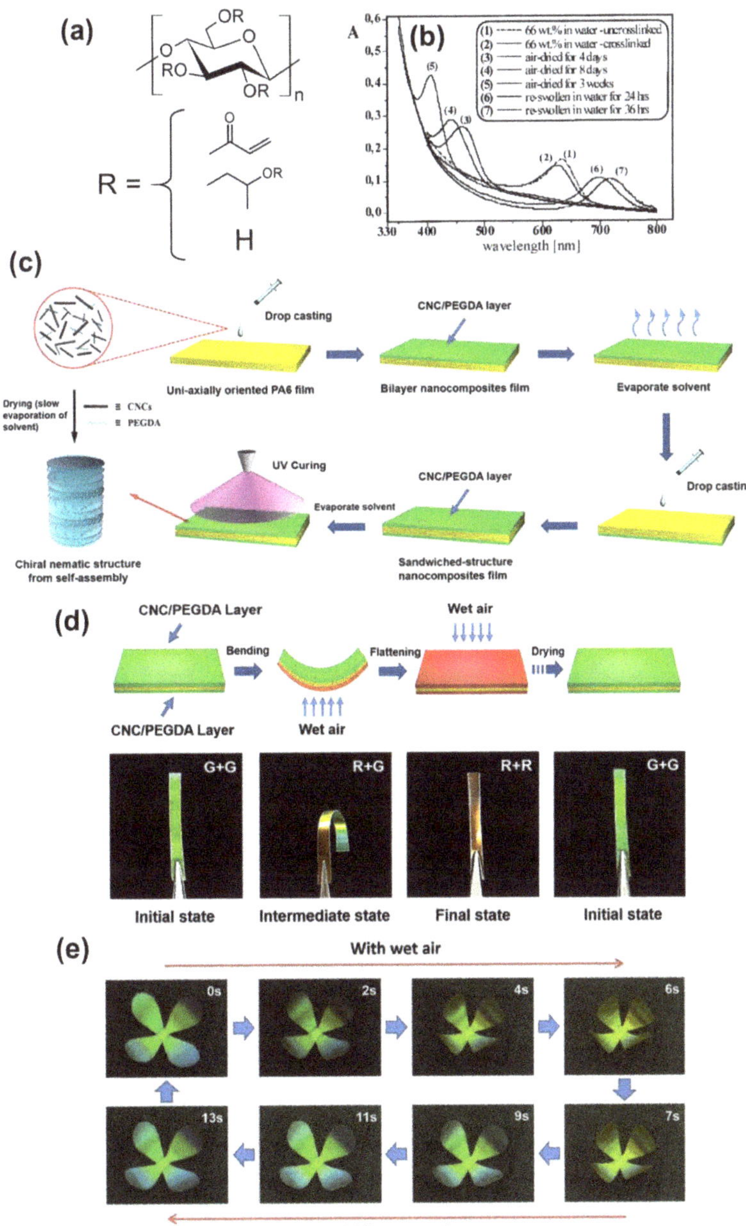

Figure 7. (**a**) Chemical structure of hydroxypropyl cellulose (HPC) acrylate and (**b**) effect of swelling and drying on UV–Vis spectra of an HPC-based cholesteric liquid crystal (ChLC) hydrogel (Reprinted from [51] by permission of Wiley). (**c**) Schematic of cellulose nanocrystal (CNC) ChLC gel/polyamide-6 (PA-6) film sandwich composites. PEGDA, polyethylene glycol diacrylate. (**d**) Schematic and visual appearances of water-vapor-induced color change as well as actuation of the sandwich composite. R: red; G: green. (**e**) Wet-air response of an artificial flower prepared from a sandwich composite (**c–e**) were reproduced from [71] by permission of The Royal Society of Chemistry.

Wu et al. [71] evaluated the humidity response of a specimen cut from the composite with a CNC/PEGDA mass ratio of 54.5/45.5 along the orientational axis of the PA-6 film. They placed the sample in a 10-mL bottle with water, and provided wet air to the sample by heating the bottle to 75 °C. The humidity-response tests indicate that the color of the side of the sample exposed to wet air redshifted, whereas the other side exhibited no color change (Figure 7d). In addition, the sample exhibited a bending deformation in response to the humid environment. These color-change and deformation phenomena are attributable to asymmetric swelling of the CNC-based ChLC gel outer layers, originating from a blockage of the water vapor by the hydrophobic PA-6 interlayer film. The researchers flattened the composite by wetting both sides and observed the original color after drying. They also prepared an artifact mimicking a four-petal flower from the sandwich composite and demonstrated the composite's humidity-responsive behavior (Figure 7e). Providing wet air to the flower mimic gave rise to folding due to the bending force of the petals, whereas removing the moisture source caused recovery of the original shape by water desorption. A composite material expressing humidity-responsive color change as well as three-dimensional deformation is attractive as a potential application in, for example, humidity sensors or wet air-driven actuators.

The humidity response of CNC-based ChLC gels is an active area of research [73–77]. The substantial hydrophilicity and useful coloration properties of CNC-based ChLCs are suitable for achieving a humidity response.

2.2.3. Colored Gels with Tailored Stimuli-Responsivity

Kelly et al. [72] produced stimuli-responsive photonic hydrogels incorporating a CNC ChLC structure. They concentrated 3 wt% CNC aqueous dispersions containing the monomers, crosslinker, and photoinitiator 2,2-diethoxyacetophenone to >60 wt% CNC, to obtain ChLC dispersions with visible color through EISA. The researchers then prepared photonic hydrogels by in situ photopolymerization of the monomer in the colored ChLC dispersions for 1 h. They used several types of monomer and crosslinker for preparing the hydrogels, such as acrylamide, *N*-isopropylacrylamide, polyethylene glycol methacrylate, *N*,*N'*-methylenebisacrylamide, and polyethylene glycol dimethacrylate.

The color of the CNC/polyacrylamide hydrogels was red-shifted after swelling in water because the CNC was diluted inside the gels (Figure 8a). Immersing the hydrogels in water corresponded to rapid swelling, reaching the swelling equilibrium after ca. 150 s (Figure 8b). Hydrogels briefly subjected to photopolymerization exhibited faster swelling than those that were sufficiently irradiated because of the loose polymer network in the former. Taking advantage of this property (controllable swelling rate), Kelly et al. [72] spatially modulated the swelling behavior by masking a section of the hydrogels during photopolymerization. Such photonic hydrogels can be used to produce a security device that exhibits a latent image by swelling in water (Figure 8c).

Because diverse monomers can be used to prepare CNC-based ChLC hydrogels, a wide range of functionality can be imparted to the hydrogels by simply changing the monomer species in the aqueous dispersions. To demonstrate this functionalization strategy, Kelly et al. [72] prepared hydrogels composed of CNC and poly(acrylic acid) (PAA), and demonstrated a pH-responsive color change originating from pH-dependent ionization of PAA (Figure 8d).

2.3. Cellulosic ChLC Mesoporous Materials

Constructing a mesoporous structure can impart valuable physical properties to cellulosic ChLC materials, such as flexibility and substantial solvent absorption. Furthermore, chiral mesoporous structures will facilitate fabrication of unprecedented functional materials, such as chiral reaction fields and optical resolution columns.

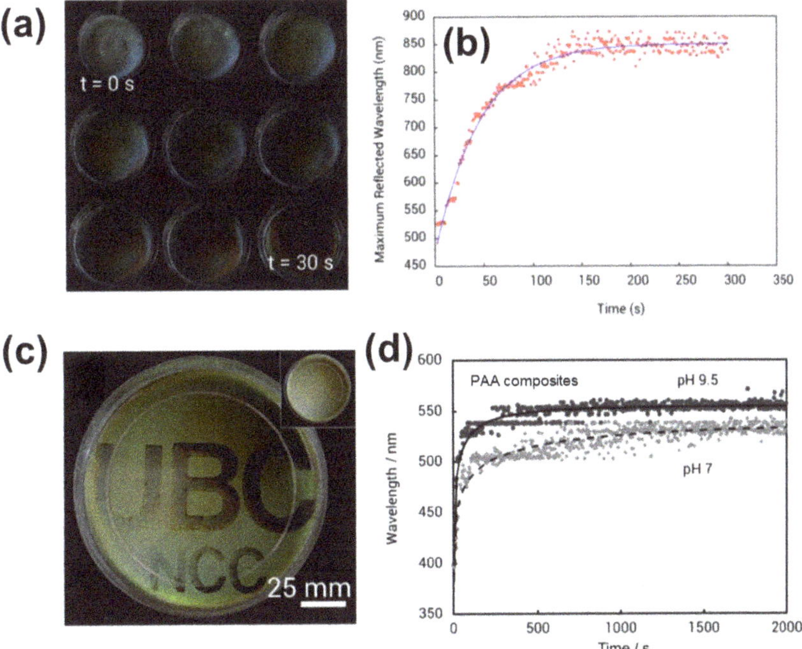

Figure 8. (**a**) Visual appearance of a cellulose nanocrystal (CNC)/polyacrylamide (PAAm) cholesteric liquid crystal (ChLC) hydrogel swollen in water for 30 s. (**b**) Time dependence of the selective reflection wavelength of a CNC/PAAm ChLC hydrogel swollen in water. (**c**) Swollen state in water for a CNC/PAAm ChLC hydrogel prepared by in situ photopolymerization of the precursor ChLC dispersion with a photomask. (**d**) Time dependence of the selective reflection wavelength of a CNC/poly(acrylic acid) (PAA) ChLC hydrogel swollen in water at pH 7 and 9.5 (**a**–**d**) were reproduced from [72] by permission of Wiley).

2.3.1. Flexible and Solvent-Responsive Mesoporous Films

Khan et al. [78] produced flexible mesoporous ChLC films by compounding CNC with phenol–formaldehyde (PF) resin. In a typical procedure, they mixed a 3.5 wt% CNC aqueous dispersion with 35 wt% PF solution in water/methanol, and then dried the mixture under ambient conditions for EISA. The researchers heated the obtained films at 75 °C for 24 h to cure the PF resin, resulting in CNC/PF resin composite films. They prepared mesoporous films upon treatment of the composites with an alkali solution or an alkali/urea mixture to remove the internal CNC, followed by supercritical drying with CO_2 (Figure 9a).

Khan et al. [78] retained coloration of the films, derived from selective reflection of CPL, even after removal of the CNCs. This retained coloration indicates that the researchers obtained a CNC-based ChLC morphology on the mesoporous structure. In addition, the mesoporous ChLC films were pliable and readily bent without structural damage (Figure 9b), in contrast to the brittle nature of composites before removing the CNCs. Tensile stress–strain curves indicate that the mesoporous films exhibited higher flexibility (lower elastic modulus) and toughness (higher elongation at break) than the corresponding nonporous films (Figure 9c). The researchers also examined the solvent-responsivity of the mesoporous ChLC films. The reflection color of the film red-shifted upon immersing the film in a water/ethanol mixed solvent, and the color corresponded to the water/ethanol ratio (Figure 9d). This is attributable to an increase in the cholesteric helical pitch originating from swelling of the mesoporous structure (Figure 9e). The color change was reversible by deswelling. Such CNC-based mesoporous ChLC films may find use as e.g., structurally colored plastics and solvent sensors.

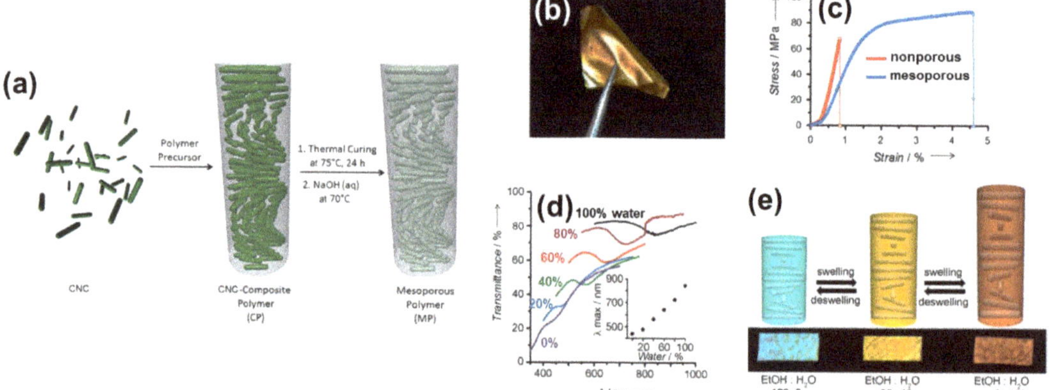

Figure 9. (**a**) Schematic for preparing mesoporous cholesteric liquid crystal (ChLC) plastics from cellulose nanocrystal (CNC)/phenol–formaldehyde (PF) resin ChLC dispersions. (**b**) Visual appearance of a mesoporous ChLC plastic deformed by bending. (**c**) Stress–strain curves of a nonporous CNC/PF resin composite film and a mesoporous ChLC plastic. (**d**) Ultraviolet–visible transmission spectra of mesoporous ChLC plastics immersed in a water/ethanol mixture. (**e**) Schematic of the swelling behavior of mesoporous ChLC plastics in water/ethanol mixtures (**a**–**e**) were reproduced from [78] by permission of Wiley).

2.3.2. Cellulose Derivative–Silica Mesoporous Hybrids for Chiral Chromatography

Sato et al. [79] prepared cellulose chlorophenylcarbamate derivative–silica hybrids with ChLC as well as mesoporous structures by a sol–gel process, and evaluated their optical and chiral resolution properties. They added 3- and 4-chlorophenyl isocyanate into 2.8 wt% cellulose in dimethylacetamide/LiCl, and carried out the reaction at various temperatures and times under nitrogen. The researchers purified the obtained cellulose 3-chlorophenylcarbamate (3Cl-CPC) and 4-chlorophenylcarbamate (4Cl-CPC; Figure 10a) by dissolution–precipitation and Soxhlet extraction. They prepared lyotropic ChLC solutions of 3Cl-CPC by dissolving 3Cl-CPC in 3-aminopropyltrimethoxysilane at 32–40 wt%, and prepared ChLC solutions of 4Cl-CPC by dissolving 4Cl-CPC in a tetramethyl orthosilicate/dimethylformamide/dichloroacetic acid mixed solvent at 32–48 wt%. The researchers exposed the ChLC solutions of 3Cl-CPC and 4Cl-CPC to air to solidify the solutions through the sol–gel process, resulting in cellulose chlorophenylcarbamate–silica hybrids.

In CD spectroscopy, the concentrated 3Cl-CPC/3-aminopropyltrimethoxysilane solutions exhibited negative peaks originating from the right-handed cholesteric helix, whereas 4Cl-CPC/tetramethyl orthosilicate/dimethylformamide/dichloroacetic acid solutions exhibited positive peaks originating from a left-handed helix (Figure 10b). Sato et al. [79] also observed the selective reflection peaks in CD spectra of silica hybrids obtained by sol–gel processing of those ChLC solutions, indicating immobilization of the ChLC structure in the hybrids (Figure 10b). The reflection peaks of the solidified samples blue-shifted compared with those of the corresponding parent solutions, likely because the samples shrank during gelation.

Sato et al. [79] attempted a chiral resolution of a racemic compound by column chromatography with the cellulose chlorophenylcarbamate–silica ChLC hybrid as the column filler. A concept of this column system is two-chirality with different scales, which are the molecular chirality of the cellulose derivative and the supramolecular chirality of the ChLC structure; whereas conventional chiral columns adopt only molecular chirality. The researchers carried out an open-column chromatographic separation of a racemate of *trans*-stilbene oxide (TSO) into two enantiomers [i.e., (*R,R*)-TSO and (*S,S*)-TSO] with a column filled with 4Cl-CPC–silica ChLC hybrid granules ($\varphi \leq \sim$120 μm). The resultant chromatogram exhibited a broad peak composed of two signals (Figure 10c), although a conventional chiral column manufactured from 4Cl-CPC enable chiral separation of TSO. The researchers attributed this insufficient chiral resolution by 4Cl-CPC–silica ChLC hybrid to fewer contacts of the filler with the chiral molecules, itself attributable to the nonuniform particle size of the filler. Technical optimization will validate the concept of two-chirality with different scales.

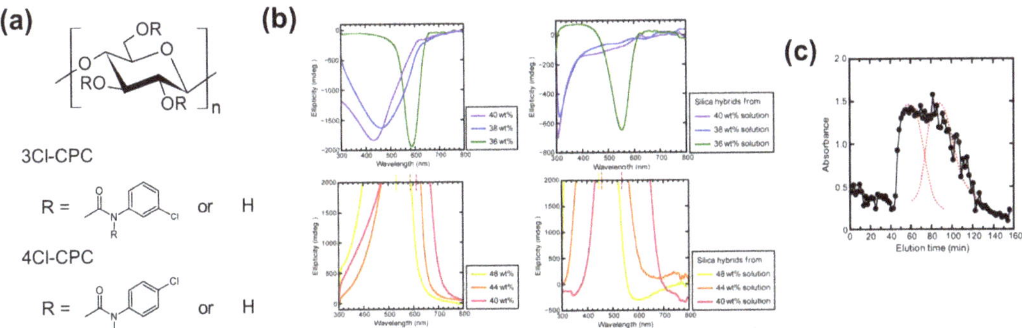

Figure 10. (**a**) Chemical structures of cellulose, 3-chlorophenylcarbamate (3Cl-CPC), and 4-chlorophenylcarbamate (4Cl-CPC). (**b**) Circular dichroism spectra of 3Cl-CPC/3-aminopropyltrimethoxysilane (left) and 4Cl-CPC/tetramethyl orthosilicate/dimethylformamide/dichloroacetic acid (right) cholesteric liquid crystal solutions with various cellulosic concentrations. (**c**) Chromatographic profile for racemic *trans*-stilbene oxide using an open column filled with a 4Cl-CPC–silica hybrid (**a**–**c**) were reproduced from [79] by permission of Elsevier).

2.4. Other Cellulosic ChLC Solid Functional Materials

There are additional reports on diverse cellulosic ChLC solid functional materials—such as structurally colored liquid marbles [80] and clay composites [81], photonic skins to visualize human motion [82,83], transistors for CPL-sensing [84], solvent-resistant photonic films [85], mechanically anisotropic composites [86], CPL-luminescent films incorporating quantum dots [87], pressure-sensitive aerogels [88], humidity-responsive photonic microarrays [89], and inkjet-printed photonic patterns [90]. These studies will facilitate expansion of the range of applications of cellulosic materials to industrial sectors in which cellulose is often considered to be incompatible with high performance.

3. Cellulosic ChLC-Based Liquid Functional Materials

3.1. Cellulosic ChLC Solutions

A representative material form of cellulosic ChLC liquid materials is solution; that is, lyotropic ChLC. Although the basic physical properties of cellulosic ChLC solutions have been thoroughly investigated to date, research on functionalization of the solutions have been limited. A core reason for this lack of research may be the high viscosity of cellulosic ChLC solutions, which leads to handling difficulties and slow stimuli-responses.

Nishio et al. [91] and Chiba et al. [92] reported cellulose derivative-based ChLC solutions doped with salts and modulated the solutions' color with an electric field. This system is a precursor of the salt-responsive color gels described in Section 2.2.1.

Nishio et al. [91] prepared ChLC solutions by dissolving HPC into aqueous salt. They accelerated dissolution of HPC by repeated centrifugation of the mixture. The selective reflection wavelength of 62.5 wt% HPC ChLC aqueous solution blue-shifted in accordance with increasing concentration of LiCl, whereas the wavelength red-shifted in accordance with increasing concentration of LiSCN (Figure 11a). This difference is likely related to the chaotropicity of the salt ions as aforementioned in Section 2.2.1. Chiba et al. [92] applied an electric field to HPC ChLC aqueous solutions containing LiSCN, and found that the positive electrode side of the solutions red-shifted and the negative electrode side blue-shifted (Figure 11b). Which electrode side exhibited a blue-shift and which side exhibited a red-shift depended on the salt species. A long time (~3 h) was necessary to reach the equilibrium solution color under the applied electric field, because of the high viscosity of the cellulosic ChLC solutions.

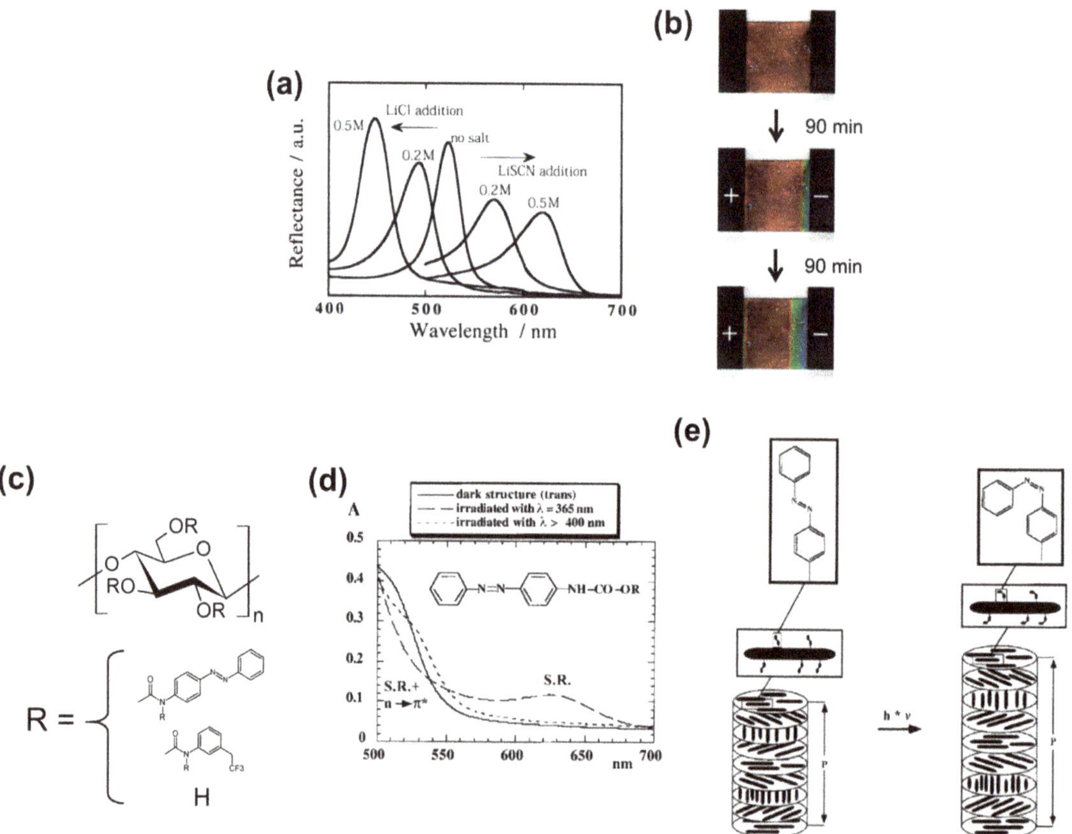

Figure 11. (**a**) Ultraviolet–visible (UV-Vis) spectra of hydroxypropyl cellulose (HPC) cholesteric liquid crystal (ChLC) aqueous solutions with various concentrations of LiCl and LiSCN. (**b**) Visual appearance of HPC ChLC aqueous solutions with LiSCN under application of an electric field (**a,b**) were reprinted with permission from [91,92]. Copyright (2021) American Chemical Society. (**c**) Chemical structure of azo-cellulose. (**d**) UV–Vis spectra of an azo-cellulose ChLC solution prior to irradiation, after irradiation at 356 nm, and after irradiation at >400 nm. (**e**) Schematic of the change in the cholesteric helical pitch of azo-cellulose ChLC solutions by irradiation (**d,e**) were reproduced from [51] by permission of Wiley).

Müller and Zentel [51] produced cellulosic ChLC solutions in which light modulates the color. To induce the photoresponsive color changes, they synthesized cellulose with two types of side groups: the trifluoromethyl phenyl carbamoyl group required for expressing ChLC, and the azobenzene side group that imparts photoresponsivity (Figure 11c).

Müller and Zentel [51] synthesized a cellulose derivative (azo-cellulose) by reacting 3-(trifluoromethyl)phenyl isocyanate and isocyanoazobenzene with cellulose in dimethylacetamide/LiCl solution at 80 °C for 4 d, followed by dissolution–precipitation with acetone and a methanol/water mixture. They determined the quantity of azobenzene side groups to be 5.4 wt% by UV–Vis spectroscopy. The researchers prepared ChLC solutions by dissolving azo-cellulose in diethylene glycol dimethacrylate at 44.3 wt%.

In the UV–Vis spectra of the azo-cellulose ChLC solution prior to irradiation, Müller and Zentel [51] observed the selective reflection peak at ca. 500 nm (Figure 11d). By irradiating the ChLC solution at 365 nm, the selective reflection peak shifted to 630 nm, and shifted back to ca. 500 nm by irradiation at 400 nm. This photoresponsive color change is attributable to light-induced *cis–trans* isomerization of the azobenzene side groups, which gives rise to a change in the cholesteric helical pitch that is attributable to the steric effect (Figure 11e). The researchers exposed the azo-cellulose ChLC solutions to irradiation for 1 h. This is indicative of a long time to reach the equilibrium of the photoresponsive color change, because of the high viscosity of the solutions.

3.2. Cellulosic ChLC Emulsions

A liquid-crystal emulsion is a system composed of dispersed liquid-crystal phase and a continuous liquid phase. Because the liquid-crystal phase is confined in a microscopic droplet, there are characteristic physical properties that are not observed in a bulk liquid-crystal. There have been many studies on liquid-crystal emulsions for applications to e.g., biosensors [93–96], omnidirectional laser emission [97–100], photonic pigments [101], and counterfeiting [102] This material form is expected to exhibit high macroscopic fluidity (in particular for molecular liquid-crystals, because polymer chains entangle only within each microdroplet) and high stimuli-responsivity originating from the large surface area of the microscopically dispersed liquid-crystal phase. Liquid-crystal emulsions therefore have great potential for liquid-type ChLC functional materials that overcome the drawbacks of lyotropic ChLCs mentioned in Section 3.1.

Li et al. [103] fabricated CNC-based ChLC emulsions by the microfluidic technique, and evaluated the particle size-induced topological transition of ChLC microdroplets. They also studied functionalization of the ChLC emulsions by loading nanoparticles (NPs) into the droplets.

For preparation of the ChLC emulsions, Li et al. [103] injected a CNC ChLC aqueous suspension as a droplet phase into the central channel of a microfluidic device, and injected a continuous phase consisting of a fluorinated oil and a copolymer surfactant into the side channels (Figure 12a). They ranged the droplet size from the micron order to hundred-micron order with low polydispersity by changing the flow rate of the oil phase.

Li et al. [103] evaluated the effect of the droplet size on the topology of the CNC ChLC droplets by POM (Figure 12b). For the larger droplets with radius (R) in the range of $40 \leq R \leq 115$ μm, they observed a concentric ring pattern with the Maltese cross. This optical pattern is attributable to the radial topology of the ChLC helices, where the helical axes of the ChLC are oriented perpendicular to the droplet surface (the pseudonematic planes are tangentially aligned to the droplet surface). The double distance between two adjacent concentric rings, corresponding to the cholesteric helical pitch, was 6.1 ± 0.3 μm; and the helical pitch in the droplets was independent on the droplet size. For the middle size of droplets, $10 \leq R \leq 40$ μm, the researchers observed a stripe pattern at the core and the aforementioned concentric ring pattern at the periphery. This stripe pattern originates from the bipolar topology of the ChLC helices, where the helical axes of the ChLC tangentially oriented to the droplet surface. For the smaller droplets, $R \leq 10$ μm, the stripe pattern accounted for almost all of the droplets. The researchers explained this size-dependent topological transition of the ChLC microdroplets in terms of a balance between the elastic energy of spherical packing of the ChLC helices and the surface anchoring energy of the pseudonematic planes.

To impart a range of functionalities to their CNC-based ChLC emulsions, Li et al. [103] loaded various NPs into the microdroplets. The inside of the droplets phase-separated into a CNC-rich ChLC region and an NP-rich isotropic region, whereby the ChLC structure was maintained even after loading the NPs. A ChLC emulsion with gold NP-loaded microdroplets exhibited extinction that is attributable to the surface plasmon resonance, and a ChLC emulsion with carbon dot-loaded microdroplets exhibited photoluminescence. In a CNC-based ChLC emulsion with magnetic NP-loaded microdroplets, the droplets moved in accordance with an applied magnetic field (Figure 12c).

Cho et al. [104] studied a system in which the dispersed ChLC microdroplets in CNC-based ChLC emulsion converted into microgels. They investigated the swelling behavior and microreactor characteristics of the ChLC microgels.

Cho et al. [104] prepared their ChLC emulsions by microfluidic emulsification of a CNC ChLC aqueous suspension containing a monomer (2-hydroxyethyl acrylate), crosslinker [poly(ethylene glycol) dimethacrylate], and photoinitiator in a similar manner as aforementioned. The precursor droplets exhibited a size-induced topological transition, which is consistent with the aforementioned microdroplets prepared from a simple CNC ChLC aqueous dispersion. The researchers then subjected the emulsions to UV-irradiation to transform the droplets into microgels by photopolymerization of the monomer and crosslinking the polymer chains.

Cho et al. [104] transferred the CNC-based ChLC microgels from oil to water, and examined the swelling behavior by POM (Figure 12d). The microgels obtained from the larger precursor droplets (126 μm) with radial topology exhibited isotropic swelling in water, giving rise to an increase in the particle size as well as a cholesteric helical pitch estimated from the distance between the adjacent concentric rings. However, the microgels prepared from smaller precursors (20 μm) with a bipolar topology anisotropically swelled, resulting in prolate microgels. There was preferential swelling along the direction perpendicular to the pseudonematic planes.

CNC-based ChLC microgels have the potential to act as microreactors because of the catalytic ability facilitated by the hydroxy groups that impart nucleophilicity, and anionic sulfate ester groups that impart cation-deposition properties. Cho et al. [104] examined the catalytic performance of the microgels for hydrolysis of 4-nitrophenyl acetate to 4-nitrophenolate, by monitoring the UV–Vis spectra of aqueous 4-nitrophenyl acetate mixed with CNC-based ChLC microgels. UV–Vis spectroscopy indicated that the intensity of the absorption peak at 270 nm (corresponding to 4-nitrophenyl acetate) decreased, whereas the intensity of the peak at 400 nm (corresponding to 4-nitrophenolate) increased, over the course of the reaction (Figure 12e). The researchers also produced Ag NPs in the microgels, where CNC acted as a reducing agent and an Ag^+ ion-absorber. The CNC–Ag NP composite ChLC microgels exhibited catalytic capability for reducing 4-nitrophenol.

Wang et al. [105] produced CNC-based ChLC emulsions and ChLC microgels. They prepared the emulsions by mixing a 4 wt% CNC aqueous suspension with acrylamide, a crosslinker, and a photoinitiator; followed by emulsification in cyclohexane with the surfactant. Moreover, the researchers exposed the ChLC emulsions to UV-irradiation for photopolymerization of the acrylamide in the microdroplets to fabricate microgels incorporating the ChLC structure. Although this work did not focus on the functionalities of the ChLC emulsions and microgel suspensions, monitoring the growth of the ChLC structure in the microdroplets by POM and direct observations of the internal ChLC structure of the microgels by SEM are noteworthy.

As described previously, emulsification of CNC-based colloidal ChLC was achieved by a microfluidic technique and simple mixing with a surfactant. However, these methods are difficult to apply to cellulose derivative-based molecular ChLC because ChLC solutions of cellulose derivatives have a higher viscosity than CNC ChLC dispersions. Therefore, other approaches are required to prepare cellulose derivative-based ChLC emulsions.

Chakrabarty et al. [106] prepared cellulose derivative-based ChLC emulsions from a cellulose derivative-*g*-block copolymer (BCP) without using a microfluidic technique or adding external surfactants. In this system, the grafted BCP-side chains underwent self-assembly in a water/oil mixture because of their amphiphilicity, resulting in ChLC emulsions containing microdroplets with an inside cellulose derivative-based ChLC solution and an outside BCP layer. The researchers evaluated the relationships between the chemical structure of cellulose derivative-*g*-BCP and the physical properties of the ChLC microdroplets.

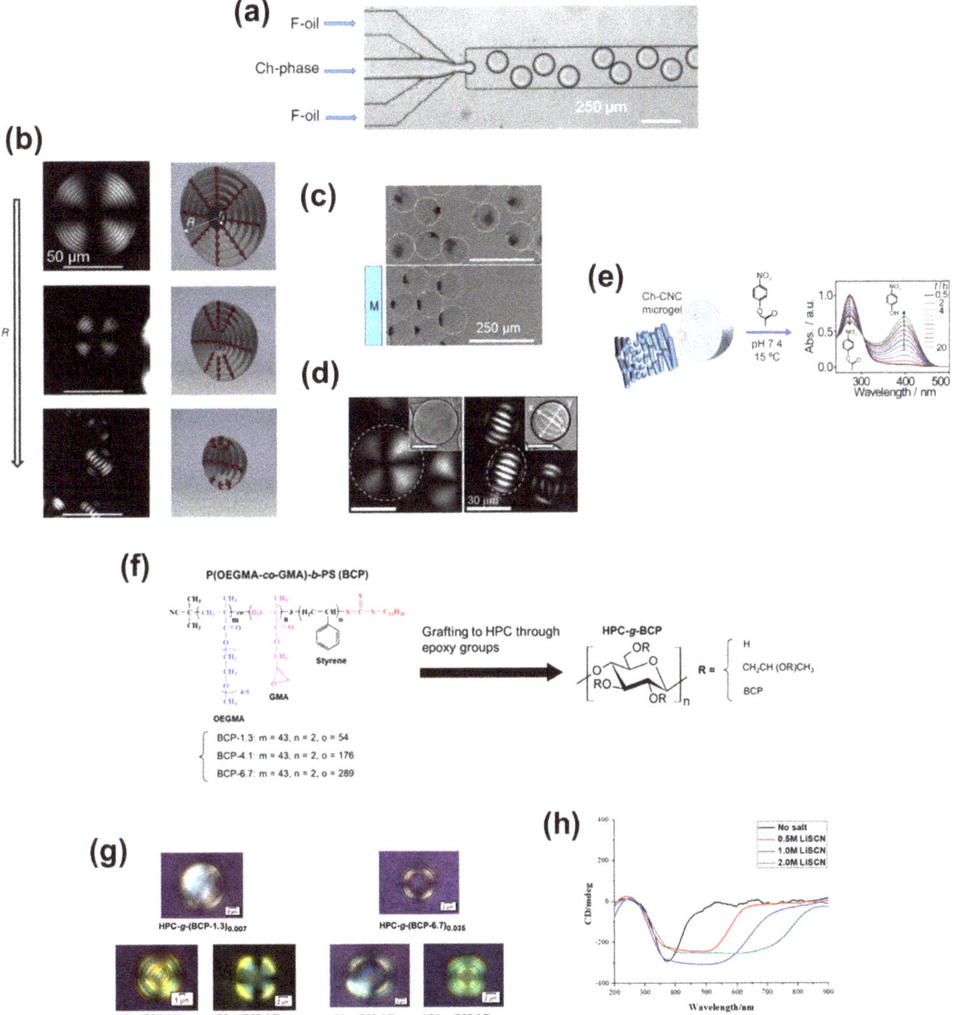

Figure 12. (**a**) Optical microscopy image of microfluidic emulsification of a cellulose nanocrystal (CNC) cholesteric liquid crystal (ChLC) aqueous dispersion in oil. (**b**) Polarized optical microscope (POM) images of CNC-based ChLC microdroplets with dimensions in the range of $40 \leq R \leq 115$ μm, $10 \leq R \leq 40$ μm, and $R \leq 10$ μm. (**c**) Optical microscopy images of CNC-based ChLC microdroplets loaded with magnetic nanoparticles with and without an applied magnetic field ((**a–c**) were reproduced from [103] by permission of Nature Springer). (**d**) POM images of CNC-based ChLC microgels in water, with insets showing the corresponding bright field images. (**e**) Time dependence of the UV–Vis spectra of aqueous 4-nitrophenyl acetate mixed with CNC-based ChLC microgels ((**d,e**) were reproduced from [104] by permission of Wiley). (**f**) Schematic for hydroxypropyl cellulose (HPC)-*g*-block copolymer (BCP) synthesis. GMA, glycidyl methacrylate; OEGMA, oligoethylene glycol methacrylate; P(OEGMA-*co*-GMA), poly(oligoethylene glycol methacrylate-*co*-glycidyl methacrylate); PS, polystyrene. (**g**) POM images of ChLC microdroplets formed by HPC-*g*-BCPs with various PS block lengths and comparable DS$_{BCP}$ (left three images), and with different DS$_{BCP}$ and comparable PS block lengths (right three images). (**h**) Circular dichroism spectra of the ChLC emulsions formed by HPC-*g*-(BCP-6.7)$_{0.006}$ with various concentrations of LiSCN ((**f–h**) were adapted with permission from [106], copyright (2021) American Chemical Society.

Chakrabarty et al. [106] synthesized a series of BCPs composed of hydrophilic poly (oligoethylene glycol methacrylate-*co*-glycidyl methacrylate) and hydrophobic polystyrene blocks by reversible addition–fragmentation chain transfer polymerization. They varied the ratio for the degree of polymerization of the hydrophilic and hydrophobic blocks (x) by changing the polystyrene block length. The researchers then grafted a series of BCPs onto HPC by ring-opening etherification involving the epoxy groups of glycidyl methacrylate and the hydroxy groups of HPC to obtain the graft copolymers HPC-*g*-(BCP-*x*)$_y$ (Figure 12f), where y represents the average number of grafted BCP chains per anhydroglucose unit (DS_{BCP}). They prepared the ChLC emulsions by adding HPC-*g*-BCPs into a water/xylene mixture and subsequent 30-min sonication. These are water-in-oil emulsions where the HPC ChLC aqueous solution is the dispersed phase and xylene is the continuous phase. The researchers determined the quantity of water such that the HPC aqueous concentration inside the droplets was 50–70 wt%, which is required to impart a visible color to the bulk system. They formed the ChLC emulsions by sonicating the mixture for only 30 min, whereas preparation of cellulose derivative-based ChLC bulk solutions requires mixing for several days.

POM observations of the ChLC emulsions, prepared from HPC-*g*-BCPs [HPC-*g*-(BCP-1.3)$_{0.007}$, HPC-*g*-(BCP-4.1)$_{0.010}$, and HPC-*g*-(BCP-6.7)$_{0.006}$] with various BCP compositions and comparable (low) DS_{BCP} values, indicate that the topological transition of the ChLC core from the radial to the bipolar occurred through an increase in the hydrophobic polystyrene length of the BCP side chains (Figure 12g). The topology of the liquid crystalline microdroplets in the emulsion was sensitive to the chemical environment at the droplet surface [93,107]. The topological transition can be interpreted as a change in the orientation state of HPC chains in response to the increase in the hydrophobicity of the droplet surface, which is imparted by the increased polystyrene block length. POM observations of ChLC emulsions, prepared from HPC-*g*-BCPs [HPC-*g*-(BCP-6.7)$_{0.035}$, HPC-*g*-(BCP-6.7)$_{0.020}$, and HPC-*g*-(BCP-6.7)$_{0.012}$] with various DS_{BCP} values and an identical BCP composition, indicate that the ChLC region inside the droplets increased in accordance with increasing DS_{BCP} (Figure 12g). These POM results indicate that the mobility of the HPC molecular chains when BCP was relatively densely grafted tended to be constrained by the shell of the droplets, rendering it difficult to supply HPC segments that would form a ChLC phase to the core.

To examine the effect of adding salt to the ChLC emulsions, Chakrabarty et al. [106] prepared emulsions from HPC-*g*-(BCP-6.7)$_{0.006}$ with LiSCN aqueous solutions at various concentrations as the water-based phase. CD spectroscopy of the ChLC emulsions with LiSCN indicated that increasing the concentration of LiSCN gave rise to a broadened CD peak of the emulsions (Figure 12h), which differed from the systematic blue-shift observed in the simple HPC aqueous solution system described in Section 3.1. This peak-broadening phenomenon is attributable to the distribution of the mobility-restricted states of the HPC chains anchored onto the BCP shell, which leads to a heterogeneous response of the cholesteric helical pitch to the salt. This unique salt-response of cellulose derivative-based ChLC emulsions can be applied to fabricate a broadband CPL filter, as revealed in Section 2.1.3.

4. Summary

We briefly described the fundamentals of cellulosic liquid crystals and reviewed research on fabricating cellulosic ChLC functional materials. As a concept of the review, we broadly organized such materials into the solid and liquid types, and further divided these in terms of their material forms, because the material form is foundational to the applications of the materials.

In cellulosic ChLC-based solid functional materials, films have been studied mostly because of the high film-formability of cellulosics. This material form is typically suitable for producing mechanochromic materials to visualize mechanical stimuli, such as compression and tensile. Large-scale manufacturing of touch-responsive cellulosic ChLC films using the roll-to-roll method is possible. Cellulosic ChLC films are also applicable as optical filters, and expansion of the filtering function has been accomplished by compounding cellulosic ChLCs with nematic liquid crystals and a surfactant. Cellulosic ChLC gels also are an active field of research. A typical function achieved by this material form is humidity-responsive color change. In particular, there are many reports regarding humidity-responsive hydrogels prepared from CNC-based ChLC aqueous dispersions because of the high hydrophilicity of CNC. Cellulosic ChLC gels can also exhibit a salt-responsive color change, and other responsivities can be imparted by compounding the gels with diverse synthetic polymers. Cellulosic ChLC-based mesoporous materials enable fabrication of structurally colored flexible plastics.

Regarding cellulosic ChLC-based liquid functional materials, studies on functionalization of cellulosic ChLC solutions have been limited; whereas fundamental aspects have been thoroughly investigated. This lack of functionalization research is likely because of the high viscosity of the solutions, leading to handling difficulties and slow stimuli-responses. The cellulosic ChLC emulsion is a material form that is expected to overcome these problems because of its high fluidity and the large surface area of the ChLC microdroplets. Although there are few reports on cellulosic ChLC emulsions, several interesting physical properties were found that were not observed in the bulk system. The topological transition of ChLC microdroplets in response to the surrounding chemical conditions will facilitate applications to chemical sensors. Because cellulosic ChLC microdroplets can load various NPs without disrupting the ChLC structure, researchers will develop useful functions by compounding cellulosic ChLCs with NPs. Furthermore, cellulosic ChLC microdroplets may serve as microreactors for chiral-selective reactions based on the chiral scaffold inside the droplets.

Cellulosic ChLC materials have the potential to exhibit diverse functions and possible applications, depending on their material forms, as presented in the current review. Whereas previous studies on cellulosic ChLC-based functional materials have generally focused on the coloration properties of cellulosic ChLCs, developing functions that take advantage of their chirality has been limited. Few researchers have produced functional materials, such as biomineralized films, conducting films, and mesoporous fillers, by using cellulosic ChLCs as a chiral scaffold; the relationship between the functionality of these materials and the ChLC structure remains to be clarified. To extend applications of cellulosic ChLC materials to more diverse fields of research, it may be important to facilitate such research. Another challenging subject on cellulosic ChLCs is the investigation of the relationship between the chemical structure of cellulosics and their liquid crystallinity, using precisely synthesized cellulosics. There has been no research on liquid crystallinity of cellulosics with regioselectively introduced side groups, as far as we know. In addition, the number of side groups on one repeating unit is intra- and intermolecularly random, for cellulosics used in previous research on cellulosic ChLCs. Controlling these structural parameters may enable excellent cellulosic ChLC-based functionalities. Accomplishing the subjects mentioned here will facilitate a more sustainable society by expanding the application scope of cellulosic materials.

Author Contributions: K.M. conceived the overall organization for this review and wrote all sections. Y.T. was consulted on the concept of the entire review and edited and revised the drafts. All authors have read and agreed to the published version of the manuscript.

Funding: The authors gratefully acknowledge the financial support for our research works on cellulosic ChLC functional materials by Grant-in-Aids for JSPS Fellows (DC2 (19J11351) to K.M.), and Scientific Research (A) (17H01480 to Y.T.) from the Japan Society for the Promotion of Science (JSPS) and the JST-Mirai Program (Grant Number JPMJMI18E3 to Y.T.) from the Japan Science and Technology Agency (JST).

Institutional Review Board Statement: Not applicable.

Informed Consent Statement: Not applicable.

Acknowledgments: We thank the editors at Edanz (https://jp.edanz.com/ac, 4 November 2021) for editing a draft of this manuscript.

Conflicts of Interest: The authors declare no conflict of interest.

References

1. Gray, D.G. Chiral nematic ordering of polysaccharides. *Carbohydr. Polym.* **1994**, *25*, 277–284. [CrossRef]
2. Lagerwall, J.P.F.; Schütz, C.; Salajkova, M.; Noh, J.; Park, J.H.; Scalia, G.; Bergström, L. Cellulose nanocrystal-based materials: From liquid crystal self-assembly and glass formation to multifunctional thin films. *NPG Asia Mater.* **2014**, *6*, 1–12. [CrossRef]
3. Schütz, C.; Bruckner, J.R.; Honorato-Rios, C.; Tosheva, Z.; Anyfantakis, M.; Lagerwall, J.P.F. From equilibrium liquid crystal formation and kinetic arrest to photonic bandgap films using suspensions of cellulose nanocrystals. *Crystals* **2020**, *10*, 199. [CrossRef]
4. Giese, M.; Blusch, L.K.; Khan, M.K.; MacLachlan, M.J. Functional materials from cellulose-derived liquid-crystal templates. *Angew. Chem. Int. Ed.* **2015**, *54*, 2888–2910. [CrossRef] [PubMed]
5. Nishio, Y.; Sato, J.; Sugimura, K. Liquid Crystals of Cellulosics: Fascinating Ordered Structures for the Design of Functional Material Systems. *Adv. Polym. Sci.* **2016**, *271*, 241–286. [CrossRef]
6. de Vries, H. Rotatory power and other optical properties of certain liquid crystals. *Acta Crystallogr.* **1951**, *4*, 219–226. [CrossRef]
7. Kuse, Y.; Asahina, D.; Nishio, Y. Molecular structure and liquid-crystalline characteristics of chitosan phenylcarbamate. *Biomacromolecules* **2009**, *10*, 166–173. [CrossRef]
8. Chakrabarty, A.; Teramoto, Y. Recent advances in nanocellulose composites with polymers: A guide for choosing partners and how to incorporate them. *Polymers* **2018**, *10*, 517. [CrossRef]
9. Beck-Candanedo, S.; Roman, M.; Gray, D.G. Effect of reaction conditions on the properties and behavior of wood cellulose nanocrystal suspensions. *Biomacromolecules* **2005**, *6*, 1048–1054. [CrossRef] [PubMed]
10. Jarvis, M.C. Structure of native cellulose microfibrils, the starting point for nanocellulose manufacture. *Philos. Trans. R. Soc. A* **2018**, *376*. [CrossRef] [PubMed]
11. Mozdyniewicz, D.J.; Nieminen, K.; Kraft, G.; Sixta, H. Degradation of viscose fibers during acidic treatment. *Cellulose* **2016**, *23*, 213–229. [CrossRef]
12. Gray, D. Recent Advances in Chiral Nematic Structure and Iridescent Color of Cellulose Nanocrystal Films. *Nanomaterials* **2016**, *6*, 213. [CrossRef] [PubMed]
13. Canejo, J.P.; Monge, N.; Echeverria, C.; Fernandes, S.N.; Godinho, M.H. Cellulosic liquid crystals for films and fibers. *Liq. Cryst. Rev.* **2017**, *5*, 86–110. [CrossRef]
14. Revol, J.F.; Bradford, H.; Giasson, J.; Marchessault, R.H.; Gray, D.G. Helicoidal self-ordering of cellulose microfibrils in aqueous suspension. *Int. J. Biol. Macromol.* **1992**, *14*, 170–172. [CrossRef]
15. Xue, M.D.; Kimura, T.; Revol, J.F.; Gray, D.G. Effects of ionic strength on the isotropic-chiral nematic phase transition of suspensions of cellulose crystallites. *Langmuir* **1996**, *12*, 2076–2082. [CrossRef]
16. Dong, X.M.; Revol, J.F.; Gray, D.G. Effect of microcrystallite preparation conditions on the formation of colloid crystals of cellulose. *Cellulose* **1998**, *5*, 19–32. [CrossRef]
17. Araki, J.; Wada, M.; Kuga, S.; Okano, T. Biréfringent glassy phase of a cellulose microcrystal suspension. *Langmuir* **2000**, *16*, 2413–2415. [CrossRef]
18. Araki, J.; Kuga, S. Effect of trace electrolyte on liquid crystal type of cellulose microcrystals. *Langmuir* **2001**, *17*, 4493–4496. [CrossRef]
19. Viet, D.; Beck-Candanedo, S.; Gray, D.G. Dispersion of cellulose nanocrystals in polar organic solvents. *Cellulose* **2007**, *14*, 109–113. [CrossRef]
20. Yi, J.; Xu, Q.; Zhang, X.; Zhang, H. Temperature-induced chiral nematic phase changes of suspensions of poly(N,N-dimethylaminoethyl methacrylate)-grafted cellulose nanocrystals. *Cellulose* **2009**, *16*, 989–997. [CrossRef]
21. Teramoto, Y. Functional thermoplastic materials from derivatives of cellulose and related structural polysaccharides. *Molecules* **2015**, *20*, 5487–5527. [CrossRef] [PubMed]
22. Werbowyj, R.S.; Gray, D.G. Liquid Crystalline Structure in Aqueous Hydroxypropyl Cellulose Solutions. *Mol. Cryst. Liq. Cryst.* **1976**, *34*, 97–103. [CrossRef]
23. Werbowyj, R.S.; Gray, D.G. Optical Properties of (Hydroxypropy)cellulose Liquid Crystals. Cholesteric Pitch and Polymer Concentration. *Macromolecules* **1984**, *17*, 1512–1520. [CrossRef]
24. Guo, J.-X.; Gray, D.G. Effect of Degree of Acetylation and Solvent on the Chiroptical Properties of Lyotropic (Acetyl)(Ethyl)Cellulose Solutions. *J. Polym. Sci. Part B Polym. Phys.* **1994**, *32*, 2529–2537. [CrossRef]
25. Guo, J.-X.; Gray, D.G. Chiroptical Behavior of (Acetyl)(ethyl)cellulose Liquid Crystalline Solutions in Chloroform. *Macromolecules* **1989**, *22*, 2086–2090. [CrossRef]
26. Siekmeyer, M.; Zugenmaier, P. Solvent dependence of lyotropic liquid-crystalline phases of cellulose tricarbanilate. *Die Makromol. Chemie* **1990**, *191*, 1177–1196. [CrossRef]
27. Ishii, H.; Sugimura, K.; Nishio, Y. Thermotropic liquid crystalline properties of (hydroxypropyl)cellulose derivatives with butyryl and heptafluorobutyryl substituents. *Cellulose* **2019**, *26*, 399–412. [CrossRef]

28. Fukuda, T.; Sugiura, M.; Takada, A.; Sato, T.; Miyamoto, T. Characteristics of Cellulosic Thermotropics. *Bull. Inst. Chem. Res. Kyoto Univ.* **1991**, *69*, 211–218.
29. Huang, B.; Ge, J.J.; Li, Y.; Hou, H. Aliphatic acid esters of (2-hydroxypropyl) cellulose-Effect of side chain length on properties of cholesteric liquid crystals. *Polymer* **2007**, *48*, 264–269. [CrossRef]
30. Nishio, Y.; Nada, T.; Hirata, T.; Fujita, S.; Sugimura, K.; Kamitakahara, H. Handedness Inversion in Chiral Nematic (Ethyl)cellulose Solutions: Effects of Substituents and Temperature. *Macromolecules* **2021**, *54*, 6014–6027. [CrossRef]
31. Osipov, M.A. Theory for cholesteric ordering in lyotropic liquid crystals. *Nuovo Cim. D* **1988**, *10*, 1249–1262. [CrossRef]
32. Habibi, Y.; Lucia, L.A.; Rojas, O.J. Cellulose Nanocrystals: Chemistry. Self-Assembly, and Applications. *Chem. Rev.* **2010**, *110*, 3479–3500. [CrossRef]
33. Tatsumi, M.; Teramoto, Y.; Nishio, Y. Polymer composites reinforced by locking-in a liquid-crystalline assembly of cellulose nanocrystallites. *Biomacromolecules* **2012**, *13*, 1584–1591. [CrossRef] [PubMed]
34. Ritcey, A.M.; Gray, D.G. Circular reflectivity from the cholesteric liquid crystalline phase of (2-ethoxypropyl)cellulose. *Macromolecules* **1988**, *21*, 1251–1255. [CrossRef]
35. Suto, S.; Suzuki, K. Crosslinked hydroxypropyl cellulose films retaining cholesteric liquid crystalline order: 2. Anisotropic swelling behaviour in water. *Polymer* **1997**, *38*, 391–396. [CrossRef]
36. Charlet, G.; Gray, D. Solid Cholesteric Films Cast from Aqueous (Hydroxypropyl)cellulose. *Macromolecules* **1987**, *20*, 33–38. [CrossRef]
37. Suto, S.; Suzuki, K. Crosslinked hydroxypropyl cellulose films retaining cholesteric liquid crystalline order. I. Effects of cast conditions and heat treatment on the textures and order of films. *J. Appl. Polym. Sci.* **1995**, *55*, 139–151. [CrossRef]
38. Nishio, Y.; Suzuki, S.; Takahashi, T. Structural Investigation of Liquid-Crystalline Ethylcellulose. *Polym. J.* **1985**, *17*, 753–760. [CrossRef]
39. Nishio, Y.; Yamane, T.; Takahashi, T. Morphological Studies of Liquid-Crystalline Cellulose Derivatives. I. Liquid-Crystalline Characteristics of Hydroxypropyl Cellulose in 2-Hydroxyethyl Methacrylate Solutions and in Polymer Composites Prepared by Bulk Polymerization. *J. Polym. Sci. Part B Polym. Phys.* **1985**, *23*, 1043–1052. [CrossRef]
40. Nishio, Y.; Yamane, T.; Takahashi, T.; Engineering, F. Morphological Studies of Liquid-Crystalline Cellulose Derivatives. II. Hydroxypropyl Cellulose Films Prepared from Liquid-Crystalline Aqueous Solutions. *J. Polym. Sci. Part B Polym. Phys.* **1985**, *23*, 1053–1064. [CrossRef]
41. Müller, M.; Zentel, R.; Keller, H. Solid Opalescent Films Originating from Urethanes of Cellulose. *Adv. Mater.* **1997**, *9*, 159–162. [CrossRef]
42. Aoki, R.; Fukawa, M.; Furumi, S. Preparation of the color films from cellulose derivatives in a diacrylate liquid. *J. Photopolym. Sci. Technol.* **2019**, *32*, 651–656. [CrossRef]
43. Shimokawa, H.; Hayata, K.; Fukawa, M.; Furumi, S. Fabrication of reflective color films from cellulose derivatives. *J. Photopolym. Sci. Technol.* **2020**, *33*, 467–471. [CrossRef]
44. Mu, X.; Gray, D.G. Formation of chiral nematic films from cellulose nanocrystal suspensions is a two-stage process. *Langmuir* **2014**, *30*, 9256–9260. [CrossRef] [PubMed]
45. Araki, J.; Wada, M.; Kuga, S.; Okano, T. Flow properties of microcrystalline cellulose suspension prepared by acid treatment of native cellulose. *Colloids Surf. A Physicochem. Eng. Asp.* **1998**, *142*, 75–82. [CrossRef]
46. Chen, G.; Hong, W. Mechanochromism of Structural-Colored Materials. *Adv. Opt. Mater.* **2020**, *8*, 1–28. [CrossRef]
47. Ito, T.; Katsura, C.; Sugimoto, H.; Nakanishi, E.; Inomata, K. Strain-responsive structural colored elastomers by fixing colloidal crystal assembly. *Langmuir* **2013**, *29*, 13951–13957. [CrossRef] [PubMed]
48. Yue, Y.; Kurokawa, T.; Haque, M.; Nakajima, T.; Nonoyama, T.; Li, X.; Kajiwara, I.; Gong, J. Mechano-actuated ultrafast full-colour switching in layered photonic hydrogels. *Nat. Commun.* **2014**, *5*, 4659. [CrossRef] [PubMed]
49. Howell, I.R.; Li, C.; Colella, N.S.; Ito, K.; Watkins, J.J. Strain-Tunable One Dimensional Photonic Crystals Based on Zirconium Dioxide/Slide-Ring Elastomer Nanocomposites for Mechanochromic Sensing. *ACS Appl. Mater. Interfaces* **2015**, *7*, 3641–3646. [CrossRef] [PubMed]
50. Lee, G.H.; Choi, T.M.; Kim, B.; Han, S.H.; Lee, J.M.; Kim, S.H. Chameleon-Inspired Mechanochromic Photonic Films Composed of Non-Close-Packed Colloidal Arrays. *ACS Nano* **2017**, *11*, 11350–11357. [CrossRef]
51. Müller, M.; Zentel, R. Cholesteric phases and films from cellulose derivatives. *Macromol. Chem. Phys.* **2000**, *201*, 2055–2063. [CrossRef]
52. Liang, H.L.; Bay, M.M.; Vadrucci, R.; Barty-King, C.H.; Peng, J.; Baumberg, J.J.; De Volder, M.F.L.; Vignolini, S. Roll-to-roll fabrication of touch-responsive cellulose photonic laminates. *Nat. Commun.* **2018**, *9*, 4632. [CrossRef]
53. Boott, C.E.; Tran, A.; Hamad, W.Y.; MacLachlan, M.J. Cellulose Nanocrystal Elastomers with Reversible Visible Color. *Angew. Chem. Int. Ed.* **2020**, *59*, 226–231. [CrossRef] [PubMed]
54. Miyagi, K.; Teramoto, Y. Dual mechanochromism of cellulosic cholesteric liquid-crystalline films: Wide-ranging colour control and circular dichroism inversion by mechanical stimulus. *J. Mater. Chem. C* **2018**, *6*, 1370–1376. [CrossRef]
55. Ritcey, A.M.; Charlet, G.; Gray, D.G. Effect of residual linear orientation on the optical properties of cholesteric films. *Can. J. Chem.* **1988**, *66*, 2229–2233. [CrossRef]
56. Miyagi, K.; Teramoto, Y. Elucidation of the Mechanism of Stress-Induced Circular Dichroic Inversion of Cellulosic/Polymer Liquid Crystalline Composites. *Macromolecules* **2020**, *53*, 3250–3254. [CrossRef]

57. Miyagi, K.; Teramoto, Y. Exploration of immobilization conditions of cellulosic lyotropic liquid crystals in monomeric solvents by in situ polymerization and achievement of dual mechanochromism at room temperature. *RSC Adv.* **2018**, *8*, 24724–24730. [CrossRef]

58. Miyagi, K.; Teramoto, Y. Function extension of dual-mechanochromism of acylated hydroxypropyl cellulose/synthetic polymer composites achieved by "moderate" compatibility as well as hydrogen bonding. *Polymer* **2019**, *174*, 150–158. [CrossRef]

59. Miyagi, K.; Teramoto, Y. Facile design of pressure-sensing color films of liquid crystalline cellulosic/synthetic polymer composites that function at desired temperatures. *Cellulose* **2019**, *26*, 9673–9685. [CrossRef]

60. Fukawa, M.; Suzuki, K.; Furumi, S. Disappearance of reflection color by photopolymerization of lyotropic cholesteric liquid crystals from cellulose derivatives. *J. Photopolym. Sci. Technol.* **2018**, *31*, 563–567. [CrossRef]

61. Arakaki, A.; Shimizu, K.; Oda, M.; Sakamoto, T.; Nishimura, T.; Kato, T. Biomineralization-inspired synthesis of functional organic/inorganic hybrid materials: Organic molecular control of self-organization of hybrids. *Org. Biomol. Chem.* **2015**, *13*, 974–989. [CrossRef]

62. Nudelman, F.; Sommerdijk, N.A.J.M. Biomineralization as an inspiration for materials chemistry. *Angew. Chem. Int. Ed.* **2012**, *51*, 6582–6596. [CrossRef]

63. Ogiwara, T.; Katsumura, A.; Sugimura, K.; Teramoto, Y.; Nishio, Y. Calcium Phosphate Mineralization in Cellulose Derivative/Poly(acrylic acid) Composites Having a Chiral Nematic Mesomorphic Structure. *Biomacromolecules* **2015**, *16*, 3959–3969. [CrossRef]

64. Katsumura, A.; Sugimura, K.; Nishio, Y. Calcium carbonate mineralization in chiral mesomorphic order-retaining ethyl cellulose/poly(acrylic acid) composite films. *Polymer* **2018**, *139*, 26–35. [CrossRef]

65. Nakao, Y.; Sugimura, K.; Nishio, Y. CaCO3 mineralization in polymer composites with cellulose nanocrystals providing a chiral nematic mesomorphic structure. *Int. J. Biol. Macromol.* **2019**, *141*, 783–791. [CrossRef]

66. De La Cruz, J.A.; Liu, Q.; Senyuk, B.; Frazier, A.W.; Peddireddy, K.; Smalyukh, I.I. Cellulose-Based Reflective Liquid Crystal Films as Optical Filters and Solar Gain Regulators. *ACS Photonics* **2018**, *5*, 2468–2477. [CrossRef]

67. Fernandes, S.N.; Almeida, P.L.; Monge, N.; Aguirre, L.E.; Reis, D.; de Oliveira, C.L.P.; Neto, A.M.F.; Pieranski, P.; Godinho, M.H. Mind the Microgap in Iridescent Cellulose Nanocrystal Films. *Adv. Mater.* **2017**, *29*. [CrossRef] [PubMed]

68. Cao, Y.; Hamad, W.Y.; MacLachlan, M.J. Broadband Circular Polarizing Film Based on Chiral Nematic Liquid Crystals. *Adv. Opt. Mater.* **2018**, *6*, 1800412. [CrossRef]

69. Lizundia, E.; Nguyen, T.D.; Vilas, J.L.; Hamad, W.Y.; MacLachlan, M.J. Chiroptical, morphological and conducting properties of chiral nematic mesoporous cellulose/polypyrrole composite films. *J. Mater. Chem. A* **2017**, *5*, 19184–19194. [CrossRef]

70. Chiba, R.; Nishio, Y.; Sato, Y.; Ohtaki, M.; Miyashita, Y. Preparation of cholesteric (hydroxypropyl) cellulose/polymer networks and ion-mediated control of their optical properties. *Biomacromolecules* **2006**, *7*, 3076–3082. [CrossRef] [PubMed]

71. Wu, T.; Li, J.; Li, J.; Ye, S.; Wei, J.; Guo, J. A bio-inspired cellulose nanocrystal-based nanocomposite photonic film with hyper-reflection and humidity-responsive actuator properties. *J. Mater. Chem. C* **2016**, *4*, 9687–9696. [CrossRef]

72. Kelly, J.A.; Shukaliak, A.M.; Cheung, C.C.Y.; Shopsowitz, K.E.; Hamad, W.Y.; MacLachlan, M.J. Responsive photonic hydrogels based on nanocrystalline cellulose. *Angew. Chem. Int. Ed.* **2013**, *52*, 8912–8916. [CrossRef] [PubMed]

73. Bumbudsanpharoke, N.; Kwon, S.; Lee, W.; Ko, S. Optical response of photonic cellulose nanocrystal film for a novel humidity indicator. *Int. J. Biol. Macromol.* **2019**, *140*, 91–97. [CrossRef] [PubMed]

74. Meng, Y.; Cao, Y.; Ji, H.; Chen, J.; He, Z.; Long, Z.; Dong, C. Fabrication of environmental humidity-responsive iridescent films with cellulose nanocrystal/polyols. *Carbohydr. Polym.* **2020**, *240*, 116281. [CrossRef]

75. Bumbudsanpharoke, N.; Lee, W.; Chung, U.; Ko, S. Study of humidity-responsive behavior in chiral nematic cellulose nanocrystal films for colorimetric response. *Cellulose* **2018**, *25*, 305–317. [CrossRef]

76. Chen, H.; Hou, A.; Zheng, C.; Tang, J.; Xie, K.; Gao, A. Light- and Humidity-Responsive Chiral Nematic Photonic Crystal Films Based on Cellulose Nanocrystals. *ACS Appl. Mater. Interfaces* **2020**, *12*, 24505–24511. [CrossRef]

77. He, Y.D.; Zhang, Z.L.; Xue, J.; Wang, X.H.; Song, F.; Wang, X.L.; Zhu, L.L.; Wang, Y.Z. Biomimetic Optical Cellulose Nanocrystal Films with Controllable Iridescent Color and Environmental Stimuli-Responsive Chromism. *ACS Appl. Mater. Interfaces* **2018**, *10*, 5805–5811. [CrossRef]

78. Khan, M.K.; Giese, M.; Yu, M.; Kelly, J.A.; Hamad, W.Y.; Maclachlan, M.J. Flexible mesoporous photonic resins with tunable chiral nematic structures. *Angew. Chem. Int. Ed.* **2013**, *52*, 8921–8924. [CrossRef]

79. Sato, J.; Sugimura, K.; Teramoto, Y.; Nishio, Y. Preparation and chiroptical properties of cellulose chlorophenylcarbamate–silica hybrids having a chiral nematic mesomorphic structure. *Polymer* **2019**, *173*, 172–181. [CrossRef]

80. Anyfantakis, M.; Jampani, V.S.R.; Kizhakidathazhath, R.; Binks, B.P.; Lagerwall, J.P.F. Responsive Photonic Liquid Marbles. *Angew. Chem. Int. Ed.* **2020**, *59*, 19260–19267. [CrossRef]

81. Trindade, A.C.; Carreto, M.; Helgesen, G.; Knudsen, K.D.; Puchtler, F.; Breu, J.; Fernandes, S.; Godinho, M.H.; Fossum, J.O. Photonic composite materials from cellulose nanorods and clay nanolayers. *Eur. Phys. J. Spec. Top.* **2020**, *229*, 2741–2755. [CrossRef]

82. Yi, H.; Lee, S.H.; Ko, H.; Lee, D.; Bae, W.G.; Kim, T.; Hwang, D.S.; Jeong, H.E. Ultra-Adaptable and Wearable Photonic Skin Based on a Shape-Memory, Responsive Cellulose Derivative. *Adv. Funct. Mater.* **2019**, *29*, 1902720. [CrossRef]

83. Zhang, Z.; Chen, Z.; Wang, Y.; Zhao, Y. Bioinspired conductive cellulose liquid-crystal hydrogels as multifunctional electrical skins. *Proc. Natl. Acad. Sci. USA* **2020**, *117*, 18310–18316. [CrossRef] [PubMed]

84. Grey, P.; Fernandes, S.N.; Gaspar, D.; Fortunato, E.; Martins, R.; Godinho, M.H.; Pereira, L. Field-Effect Transistors on Photonic Cellulose Nanocrystal Solid Electrolyte for Circular Polarized Light Sensing. *Adv. Funct. Mater.* **2019**, *29*, 1805279. [CrossRef]
85. Zhang, F.; Ge, W.; Wang, C.; Zheng, X.; Wang, D.; Zhang, X.; Wang, X.; Xue, X.; Qing, G. Highly Strong and Solvent-Resistant Cellulose Nanocrystal Photonic Films for Optical Coatings. *ACS Appl. Mater. Interfaces* **2021**, *13*, 17118–17128. [CrossRef] [PubMed]
86. Tatsumi, M.; Kimura, F.; Kimura, T.; Teramoto, Y.; Nishio, Y. Anisotropic Polymer Composites Synthesized by Immobilizing Cellulose Nanocrystal Suspensions Specifically Oriented under Magnetic Fields. *Biomacromolecules* **2014**, *15*, 4579–4589. [CrossRef]
87. Xu, M.; Ma, C.; Zhou, J.; Liu, Y.; Wu, X.; Luo, S.; Li, W.; Yu, H.; Wang, Y.; Chen, Z.; et al. Assembling semiconductor quantum dots in hierarchical photonic cellulose nanocrystal films: Circularly polarized luminescent nanomaterials as optical coding labels. *J. Mater. Chem. C* **2019**, *7*, 13794–13802. [CrossRef]
88. Cao, Y.; Lewis, L.; Hamad, W.Y.; MacLachlan, M.J. Pressure-Responsive Hierarchical Chiral Photonic Aerogels. *Adv. Mater.* **2019**, *31*, 1808186. [CrossRef]
89. Zhao, T.H.; Parker, R.M.; Williams, C.A.; Lim, K.T.P.; Frka-Petesic, B.; Vignolini, S. Printing of Responsive Photonic Cellulose Nanocrystal Microfilm Arrays. *Adv. Funct. Mater.* **2019**, *29*, 1804531. [CrossRef]
90. Chen, R.; Feng, D.; Chen, G.; Chen, X.; Hong, W. Re-Printable Chiral Photonic Paper with Invisible Patterns and Tunable Wettability. *Adv. Funct. Mater.* **2021**, *31*, 2009916. [CrossRef]
91. Nishio, Y.; Kai, T.; Kimura, N.; Oshima, K.; Suzuki, H. Controlling the selective light reflection of a cholesteric liquid crystal of (hydroxypropyl) cellulose by electrical stimulation. *Macromolecules* **1998**, *31*, 2384–2386. [CrossRef]
92. Chiba, R.; Nishio, Y.; Miyashita, Y. Electrooptical Behavior of Liquid-Crystalline (Hydroxypropyl) cellulose/Inorganic Salt Aqueous Solutions. *Macromolecules* **2003**, *36*, 1706–1712. [CrossRef]
93. Sivakumar, S.; Wark, K.L.; Gupta, J.K.; Abbott, N.L.; Caruso, F. Liquid crystal emulsions as the basis of biological sensors for the optical detection of bacteria and viruses. *Adv. Funct. Mater.* **2009**, *19*, 2260–2265. [CrossRef]
94. Khan, M.; Park, S.Y. General Liquid-crystal droplets produced by microfluidics for urea detection. *Sens. Actuators B Chem.* **2014**, *202*, 516–522. [CrossRef]
95. Lee, H.G.; Munir, S.; Park, S.Y. Cholesteric Liquid Crystal Droplets for Biosensors. *ACS Appl. Mater. Interfaces* **2016**, *8*, 26407–26417. [CrossRef]
96. Niu, X.; Luo, D.; Chen, R.; Wang, F.; Sun, X.; Dai, H. Optical biosensor based on liquid crystal droplets for detection of cholic acid. *Opt. Commun.* **2016**, *381*, 286–291. [CrossRef]
97. Uchida, Y.; Takanishi, Y.; Yamamoto, J. Controlled fabrication and photonic structure of cholesteric liquid crystalline shells. *Adv. Mater.* **2013**, *25*, 3234–3237. [CrossRef]
98. Iwai, Y.; Kaji, H.; Uchida, Y.; Nishiyama, N. Chemiluminescence emission in cholesteric liquid crystalline core–shell microcapsules. *J. Mater. Chem. C* **2014**, *2*, 4904–4908. [CrossRef]
99. Iwai, Y.; Kaji, H.; Uchida, Y.; Nishiyama, N. Temperature-dependent Color Change of Cholesteric Liquid Crystalline Core-shell Microspheres. *Mol. Cryst. Liq. Cryst.* **2015**, *615*, 9–13. [CrossRef]
100. Humar, M.; Muševič, I. 3D microlasers from self-assembled cholesteric liquid-crystal microdroplets. *Opt. Express* **2010**, *18*, 26995–27003. [CrossRef]
101. Lee, S.S.; Kim, S.K.; Won, J.C.; Kim, Y.H.; Kim, S.-H. Reconfigurable Photonic Capsules Containing Cholesteric Liquid Crystals with Planar Alignment. *Angew. Chem.* **2015**, *127*, 15481–15485. [CrossRef]
102. Schwartz, M.; Lenzini, G.; Geng, Y.; Rønne, P.B.; Ryan, P.Y.A.; Lagerwall, J.P.F. Cholesteric Liquid Crystal Shells as Enabling Material for Information-Rich Design and Architecture. *Adv. Mater.* **2018**, *30*, 1707382. [CrossRef]
103. Li, Y.; Jun-Yan Suen, J.; Prince, E.; Larin, E.M.; Klinkova, A.; Thérien-Aubin, H.; Zhu, S.; Yang, B.; Helmy, A.S.; Lavrentovich, O.D.; et al. Colloidal cholesteric liquid crystal in spherical confinement. *Nat. Commun.* **2016**, *7*, 12520. [CrossRef] [PubMed]
104. Cho, S.; Li, Y.; Seo, M.; Kumacheva, E. Nanofibrillar Stimulus-Responsive Cholesteric Microgels with Catalytic Properties. *Angew. Chem. Int. Ed.* **2016**, *55*, 14014–14018. [CrossRef] [PubMed]
105. Wang, P.X.; Hamad, W.Y.; MacLachlan, M.J. Polymer and Mesoporous Silica Microspheres with Chiral Nematic Order from Cellulose Nanocrystals. *Angew. Chem. Int. Ed.* **2016**, *55*, 12460–12464. [CrossRef] [PubMed]
106. Chakrabarty, A.; Miyagi, K.; Maiti, M.; Teramoto, Y. Topological Transition in Spontaneously Formed Cellulosic Liquid-Crystalline Microspheres in a w/o Emulsion. *Biomacromolecules* **2018**, *19*, 4650–4657. [CrossRef] [PubMed]
107. Gupta, J.K.; Zimmerman, J.S.; De Pablo, J.J.; Caruso, F.; Abbott, N.L. Characterization of adsorbate-induced ordering transitions of liquid crystals within monodisperse droplets. *Langmuir* **2009**, *25*, 9016–9024. [CrossRef]

 *nanomaterials*

Article

Nanocellulose Xerogel as Template for Transparent, Thick, Flame-Retardant Polymer Nanocomposites

Wataru Sakuma [1,*], Shuji Fujisawa [1], Lars A. Berglund [2] and Tsuguyuki Saito [1,*]

1   Department of Biomaterial Sciences, Graduate School of Agricultural and Life Sciences,
    The University of Tokyo, 1-1-1 Yayoi, Bunkyo-ku, Tokyo 113-8657, Japan; afujisawa@g.ecc.u-tokyo.ac.jp
2   Department of Fibre and Polymer Technology, Wallenberg Wood Science Center,
    KTH Royal Institute of Technology, SE-100 44 Stockholm, Sweden; blund@kth.se
*   Correspondence: sakuma.wataru@gmail.com (W.S.); saitot@g.ecc.u-tokyo.ac.jp (T.S.);
    Tel.: +81-3-5841-5271 (T.S.)

Abstract: Cellulose nanofibers (CNFs) have excellent properties, such as high strength, high specific surface areas (SSA), and low coefficients of thermal expansion (CTE), making them a promising candidate for bio-based reinforcing fillers of polymers. A challenge in the field of CNF-reinforced composite research is to produce strong and transparent CNF/polymer composites that are sufficiently thick for use as load-bearing structural materials. In this study, we successfully prepared millimeter-thick, transparent CNF/polymer composites using CNF xerogels, with high porosity (~70%) and high SSA (~350 m^2 g^{-1}), as a template for monomer impregnation. A methacrylate was used as the monomer and was cured by UV irradiation after impregnation into the CNF xerogels. The CNF xerogels effectively reinforced the methacrylate polymer matrix, resulting in an improvement in the flexural modulus (up to 546%) and a reduction in the CTE value (up to 78%) while maintaining the optical transparency of the matrix polymer. Interestingly, the composites exhibited flame retardancy at high CNF loading. These unique features highlight the applicability of CNF xerogels as a reinforcing template for producing multifunctional and load-bearing polymer composites.

Keywords: cellulose nanofibers; nanocomposite; xerogel; flame-retardant

Citation: Sakuma, W.; Fujisawa, S.; Berglund, L.A.; Saito, T. Nanocellulose Xerogel as Template for Transparent, Thick, Flame-Retardant Polymer Nanocomposites. *Nanomaterials* **2021**, *11*, 3032. https://doi.org/10.3390/nano11113032

Academic Editor: Wei Zhang

Received: 21 October 2021
Accepted: 9 November 2021
Published: 12 November 2021

Publisher's Note: MDPI stays neutral with regard to jurisdictional claims in published maps and institutional affiliations.

1. Introduction

Polymers reinforced with stiff fillers are candidates for use as low-density structural materials. The effect of reinforcement can be enhanced by nanoscale reinforcement with a large interfacial area between the polymer matrix and reinforcement [1]. Cellulose nanofibers (CNF) are representative nanoscale bio-based reinforcements with excellent mechanical properties and thermal dimensional stability [2]. Therefore, the production of CNF composites with various polymer matrices through diverse preparation processes has been explored in past research studies [2–4].

To maximize the potential of CNF composites as structural materials, the following two points are required: (1) high CNF content in the polymer matrix without aggregation, and (2) sufficient thickness, to enable support of high loads. In the field of cellulose nanofiber science, solvent casting is commonly used for the preparation of CNF composites. This process offers transparent, homogeneous, and high-CNF-content polymer composites [4,5]. However, the resulting composites are typically in the form of thin films with thicknesses of around 100 μm or less. To overcome this drawback in solvent casting, the lamination of such thin films using polymers has been proposed [6]. Although the laminated composites were highly transparent and had excellent mechanical properties, the mechanical properties decreased as the number of laminated films increased. Another method of preparing CNF composites is melt compounding, which is practically used and is scalable [7]. Despite these advantages, the addition of CNF leads to increases in the viscosity of the molten polymer, which have been an obstacle to achieving a high CNF content [8]. In addition, in the melt-compounding process, the low dispersibility of CNFs in molten polymers has to be overcome. To improve their dispersibility, CNFs were commonly subjected to surface modification to alter their hydrophilic nature to hydrophobic. However, this modification often decreased the strength of the CNF network by prohibiting strong interactions between the CNFs [9].

One solution to these problems is to impregnate the monomer into the CNF network, followed by curing of the monomer [4,8,10–13]. Through this method, CNF-rich composites with a highly dispersed CNF network can be obtained without surface modification. In previous studies, thin CNF sheets or delignified wood structures were used as CNF networks [14,15]. The former, being thin materials, cannot be used as structural materials, whereas the latter have a low specific surface area (SSA) and thus cannot maximize the potential of CNFs.

Recently, we developed optically transmissive mesoporous CNF xerogels with high porosity (70–80%) and high SSA (>350 m^2 g^{-1}) [16,17]. Xerogels are porous materials produced through the ambient pressure drying of wet gels. Due to this scalable drying process, xerogels with thicknesses of several millimeters were obtained. Moreover, the CNF xerogels combined high stiffness and high SSA, making them ideal for use as a template for preparing strong and transparent CNF composites.

This study was thus aimed at preparing strong, transparent, thick CNF-rich polymer composites from CNF xerogels using an impregnation method. The CNF content in the composite was varied within a range of 30–80 vol% via the simple densification of CNF xerogels prior to monomer impregnation.

2. Materials and Methods

2.1. Materials and Chemicals

TEMPO-oxidized pulp with a carboxylate content of approximately 1.8 mmol g^{-1}, which was kindly provided by DKS Co. Ltd., (Kyoto, Japan) was used as the starting material for the CNFs (see Figure S1a for Fourier transform infrared (FTIR) spectroscopy analysis) [18]. Aluminum chloride hexahydrate was purchased from FUJIFILM Wako Pure Chemical Corporation (Osaka, Japan). 1-hydroxycyclohexyl phenyl ketone (UV-sensitive radical initiator) and 1,6-hexanediol dimethacrylate were purchased from Tokyo Chemical Industry Co., Ltd. (Tokyo, Japan). All chemicals were used as received.

2.2. Preparation of CNF Xerogels

The CNF xerogel was prepared in accordance with our previously reported procedure (Figure 1) [16,17]. A CNF water dispersion was prepared via mechanical disintegration of the TEMPO-oxidized pulp using a high-pressure water jet system (HJP-25005X, Sugino Machine Limited, Toyama, Japan). The width and length of the dispersed CNFs were approximately 2–3 and 300–500 nm, respectively (see Figure S1b for atomic force microscopy (AFM) image). The dispersion was then concentrated to 1.0 wt%. A 0.1 M $AlCl_3$ solution was dropped onto the dispersion to obtain the CNF hydrogels. After the solvent of the hydrogels was exchanged with ethanol and hexane, the wet gels were evaporated under ambient pressure at room temperature. The porosity and SSA of the xerogels were 70% and 350 m^2 g^{-1}, respectively. After the xerogels were processed into a certain dimension via sawing and polishing, the CNF content in the final composites was adjusted to be between 30 and 80 vol% via the uniaxial compression of the xerogels. The xerogels were dried at 70 °C for 3 h under reduced pressure prior to the following impregnation procedures.

2.3. Preparation of CNF Composites

The initiator was mixed with the monomer at 0.5 wt% for 10 min. After nitrogen purging of the mixture for 5 min, the xerogels were dipped into the monomer solution and then placed under reduced pressure (<1 Pa) until the bubbles arising from the xerogels disappeared. The monomer-containing xerogels were sandwiched between PET films (250 μm thick) with a silicone rubber spacer, and the set was sandwiched between glass plates. Each side of the specimen was then subjected to UV curing for 90 s (for a total of 3 min per specimen) using a high-pressure UV lamp unit (OPM2-502HQ, Ushio Inc., Tokyo, Japan). A pristine polymer matrix (denoted as 0 vol% CNF) was prepared using the same protocol as that used for the composites. The specimens were conditioned at 23 °C and 50% relative humidity for at least 1 d before use.

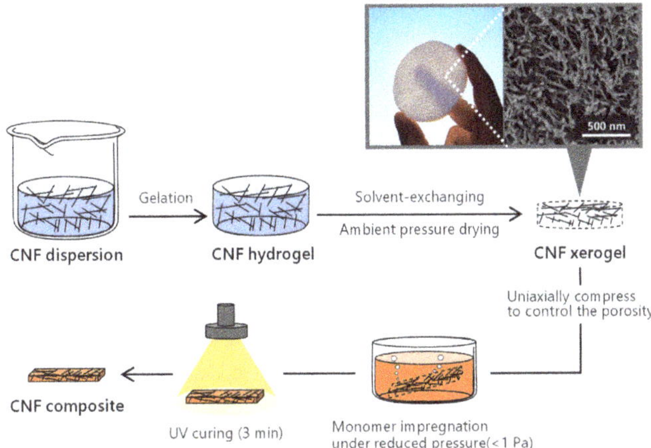

Figure 1. Scheme of CNF composite preparation.

2.4. Analysis

FTIR spectrum of the CNFs was collected using a FT/IR-6100 (JASCO Corp., Tokyo, Japan). AFM observation of the CNFs was conducted using a MultiMode 8 microscope (Bruker, Billerica, MA, USA) equipped with a NanoScope V controller. Diluted CNF water dispersion (0.0005 wt%) was dropped onto a mica plate, and the dried plate was used for the observation. The total light transmittance of the specimens was measured using a UV-Vis V670 (JASCO Corp., Tokyo, Japan) equipped with a horizontal sampling integrating sphere unit (PIN-757). The haze values of the specimens were calculated according to the ASTM D1003 "Standard Test Method for Haze and Luminous Transmittance of Transparent Plastics" [15,19]. The transmittances of composites with different thicknesses were fitted using the following equation, which describes the transmittance of a material at a certain refractive index and wavelength [20]:

$$T = (1 - R)^2 \cdot exp(-\alpha x),\tag{1}$$

where R, α, and x are the coefficient of reflection, attenuation coefficient, and specimen thickness, respectively. The α value represents the intensity of attenuation; the higher the α, the more intense the attenuation as the thickness increases. The fitting was carried out with varying values of R and α. A three-point bending test was performed using a EZ-SX (Shimadzu Corp., Kyoto, Japan) equipped with a 500 N load cell. The dimensions of the specimens, span, and crosshead speed were set to $1 \times 5 \times 30$ mm³, 20 mm, and 5 mm min⁻¹, respectively. Cross-sectional scanning electron microscopy (SEM) observations of the composites were performed using a S-4800 field emission microscope (Hitachi Ltd., Tokyo, Japan) at 1 kV. The specimens were pretreated with a Neo Osmium Coater (Meiwafosis Co., Ltd., Tokyo, Japan) at 5 mA for 10 s. X-ray diffraction measurements were conducted using a MicroMax-007 HF (Rigaku Corp., Tokyo, Japan). The degree of CNF orientation was calculated from the azimuthal profile of 200 reflections [21]. The thermal expansion behaviors of the composites were analyzed using a TMA-60 (Shimadzu Corp., Kyoto, Japan) with a load of 0.03 N and a heating ratio of 2 °C min⁻¹ under a nitrogen atmosphere. The dimensions of the specimens and the span were set to $1 \times 5 \times 12$ mm³ and 10 mm, respectively. Before measurement, the specimens were dried in the instrument at 120 °C for 90 min. The coefficients of thermal expansion (CTE) of the composites at 30–120 °C were calculated for the thermal expansion curves. The thermal degradation of the composite was investigated using a TA-50 (Shimadzu Corp., Kyoto, Japan) at a heating rate of 10 °C min^{-1} up to 500 °C under a nitrogen or artificial air atmosphere. The starting points of the thermogravimetric (TG) curves were shifted to the value at 100 °C to ignore the weight loss due to humidity in the composites. To investigate the flame retardancy of the composites, the edges of the specimens were exposed to a flame for 3 s and then for 6 s. The dimensions of the specimens were $1 \times 5 \times 20$ mm^3. All experiments were conducted at least 3 times for each condition.

3. Results and Discussion

3.1. Preparation of CNF Composite

Translucent CNF xerogels became highly transparent after impregnation and curing of the monomer (Figure 2a and Figure S1c). The match of reflective indices between the polymer and CNFs suppressed light scattering [15]. All of the CNF composites exhibited clear birefringence when observed between crossed polarizers, because of the optical anisotropy of the CNFs (Figure 2b) [22]; crystalline CNFs were present throughout the amorphous polymer matrix. The scalable preparation process of the xerogels enabled the production of a thick composite (Figure 2c). To the best of our knowledge, this is the first ever successful one-shot production of a several-millimeter-thick transparent CNF-reinforced polymer composite. The density of the composite exhibited a roughly linear dependence on the CNF content, but the value of the pristine polymer was slightly lower than the extrapolated value at 0 vol% (Figure 2d). This result suggests that the density of the polymer matrix varied based on the presence or absence of a CNF network. The extrapolated value at 100 vol% was also slightly lower than the density of the CNF skeleton (1.58 g cm^{-3}). Therefore, limited air voids of up to 9 vol% is one possibility, or increased free volume in the polymer matrix.

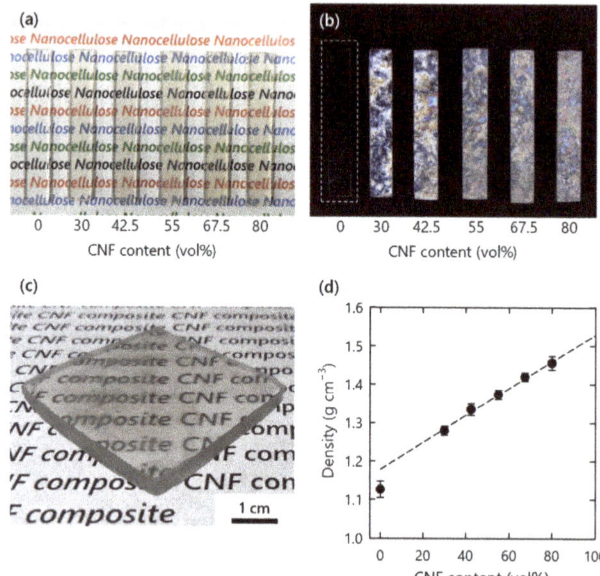

Figure 2. (**a**) Appearance of CNF composites (1 mm thickness); (**b**) appearance of composites observed between crossed polarizers; (**c**) large CNF composite (30 vol%) with 4 mm thickness; (**d**) densities of composites with different CNF contents.

3.2. Optical Properties

The optical properties of the composite were investigated using UV-Vis spectroscopy. As mentioned in the previous section, all the composites showed high transparencies in the visible-light range, regardless of the CNF content (Figure 3a). At a CNF content of 50–70 vol%, the total transmittance and haze value at a wavelength of 600 nm exhibited their lowest and highest values, respectively (Figure 3b). This phenomenon is likely explained by light scattering at the interface between the polymer matrix and the CNF. The number of scattering interfaces should increase as the CNF content increases. However, over a certain threshold of the CNF content, CNFs should start to contact with one another. Accordingly, the scattering interfaces between the polymer matrix and the CNFs are supposed to decrease at such a high CNF content [23]. Therefore, we assumed that the number of scattering interfaces reached its maximum value at a CNF content of 50–70 vol% and then decreased in the CNF content from 70 to 80 vol%, resulting in an increase in transmittance and a decrease in haze value. It should be emphasized that the reduction in transmittance relative to that of the pristine polymer matrix was only 10% for the composite with a high CNF content of 80 vol% and a thickness of 1 mm.

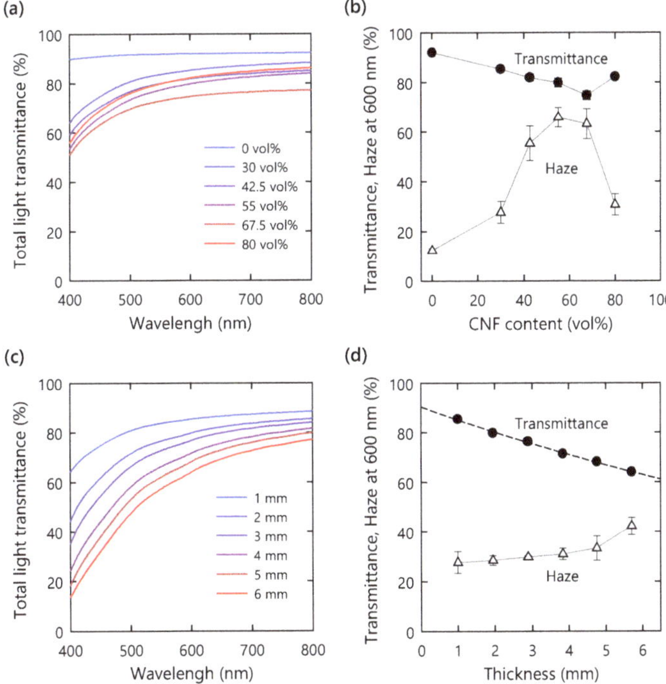

Figure 3. (a) Total light transmittance spectra of CNF composites with different CNF contents at 1 mm thickness; (b) transmittance and haze values at wavelength of 600 nm as a function of CNF content, same thickness; (c) total light transmittance spectra of composites (30 vol%) with different thicknesses; (d) transmittance and haze values at wavelength of 600 nm as functions of specimen thickness.

To examine the relationship of the optical properties with the thickness, 30 vol% CNF composites with different thicknesses were prepared. As the thickness was increased, the total transmittance and haze value of the composites gradually decreased and increased, respectively (Figure 3c,d). The transmittances at different wavelengths were fitted using Equation (1) (Figure 4a). At a wavelength of 600 nm, the α value in Equation (1) was calculated to be 0.06 mm^{-1}, which is lower than the previously reported value for a delignified-wood-based polymer composite without surface modification (0.16 mm^{-1}) [24]. The low dependence of the optical properties on thickness implies the absence of micron-sized voids in the composite and the favorable match in refractive index between the polymer and the CNFs [24].

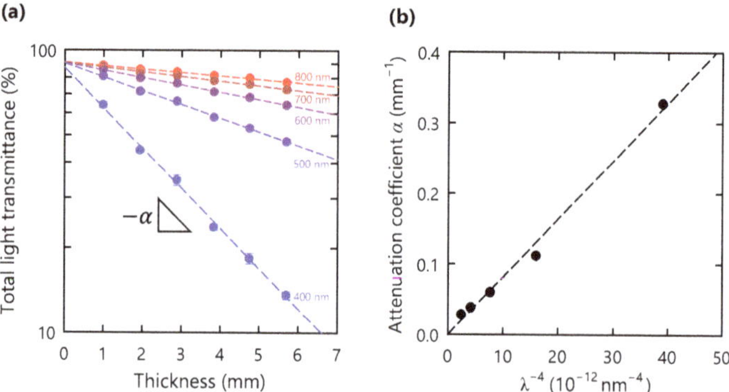

Figure 4. (**a**) Total light transmittance of CNF composites (30 vol%) with different specimen thicknesses at different wavelengths; (**b**) attenuation coefficient (α) plot as function of inverse fourth power of wavelength (λ^{-4}).

The α value has a proportional relationship with the number of scattering centers per unit volume (N), the radius of scattering centers (r), and wavelength (λ) as follows [20]:

$$\alpha = A \cdot N \cdot r^6 \cdot \lambda^{-4}, \tag{2}$$

where A is a constant. This simplified equation for Rayleigh scattering involves the following assumptions: (1) negligible light absorption; (2) the constant refractive index of the polymer and CNF as a function of wavelength; and (3) a much smaller r value compared to the wavelength. Interestingly, the α values of our composites exhibited a nearly linear dependence on λ^{-4} in the wavelength range of 400–800 nm (Figure 4b). Therefore, the number of scattering centers per unit volume (N) and radius (r) of centers (i.e., CNFs in the composite) are constants in Equation (2). These results indicate that the CNF composites have a highly homogeneous structure throughout the material with a thickness of up to 6 mm.

3.3. Mechanical Properties

The mechanical properties of the composites were evaluated using a three-point bending test. The CNF network in the polymer exhibited a significant reinforcement effect (Figure 5a). The addition of 30 vol% CNF doubled the modulus and strength of the pristine polymer. Furthermore, the composite with 80 vol% CNF reached a strength of 160 MPa, which is comparable to those of glasses [25]. The strain at failure of the composite was slightly lower than that of the pristine polymer, although the matrix also has low strain to failure. The moduli of the composites exhibited a linear relationship with CNF content (Figure 5b). A simple mixing model was used [26]:

$$E_c = E_{np} V_f + E_m (1 - V_f) \tag{3}$$

where E_c, E_m, and E_{np} are the elastic moduli of the composite, polymer matrix, and neat CNF nanopaper, respectively, and V_f is the volume fraction of CNF (note that the CNF nanopaper refers to a film obtained through the evaporative drying of CNF water dispersion). E_{np}, or effective reinforcement modulus, was set to be a free parameter and estimated from linear regression of the experimental data. The obtained E_{np} for the reinforcement phase, 11.3 GPa, was within the range of that of neat, dense CNF networks (9–15 GPa) [21,26], suggesting that the network structure of the xerogel has a dominant role in the stiffness of the composites [10,26]. One of the challenges in the field of CNF composites is the decrease in the reinforcement efficiency at high fiber content due to CNF aggregation [8]. This study avoided critical aggregation and achieved a linear reinforcement effect in a wide range of CNF content. The values of flexural strength and work of fracture (area under stress–strain curve) were very scattered, probably because of technical problems in the composite preparation, such as voids in the composites (Figure S2a,b). Nonetheless, improvements in the strength and work of fracture due to CNF addition were still confirmed.

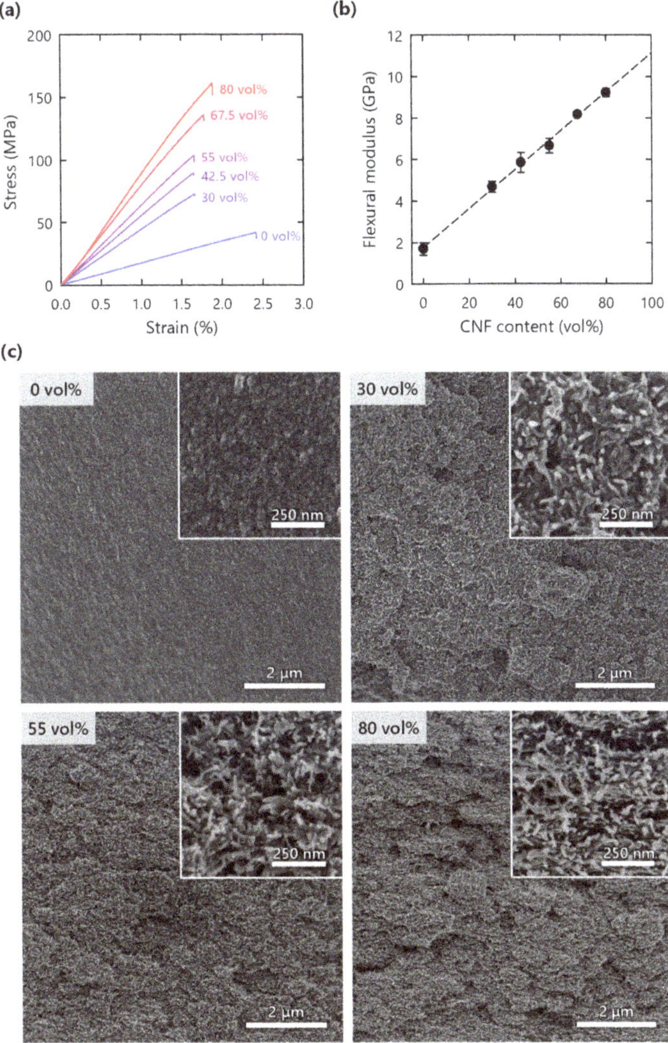

Figure 5. (a) Stress–strain curves estimated from three-point bending test of CNF composites; (b) flexural modulus of composite as function of CNF content; (c) SEM images of fractured surfaces of composites.

Through SEM observations of the fractured surface, a macroscopically rough structure was confirmed for the CNF composite, whereas a smooth fracture surface was observed for the pristine polymer (Figures 5c and S2c). These morphologies suggest more complex crack propagation due to the CNFs [27,28]. This consumes more energy and may contribute to an increase in work of fracture. A homogeneous CNF network was observed in the composites even at a high CNF content, without any sign of layered structures (inset of Figures 5c and S2c). In addition, the individual fibers appear to be uncovered by the polymer, implying pull-out of the CNFs although the lengths are short.

These results indicate that the mechanical properties of the composites were governed by the CNF reinforcement network.

3.4. Thermal Expansion Behavior

In the case of thick and large structural materials, dimensional changes due to thermal expansion are more pronounced. In this study, the nanoscale CNF network successfully restricted the expansion of the polymer matrix in thick composites (Figure 6a). The CTE values, which were calculated from the slopes of the thermal expansion curves, decreased as the CNF content increased (Figure 6b). The composite with 80 vol% CNF exhibited a 78% reduction in the CTE value with respect to that of the pristine polymer. Meanwhile, a number of previous studies on thin-film CNF composites with low CNF content reported comparable or even lower CTE values [5,29–31]. Such low CTE values for thin-film composites may have been achieved through strong in-plane orientations of CNFs in the composites [32,33]. By comparison, in this study, the degree of CNF in-plane orientation in the composite was not as high as those of thin-film composites (Figure S3).

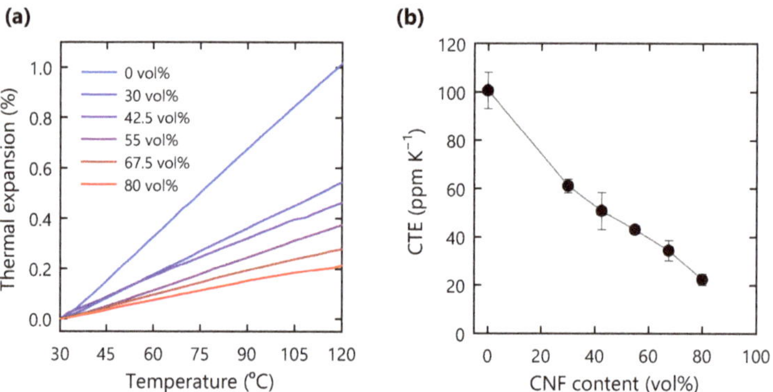

Figure 6. (a) Thermal expansion behavior of CNF composites; (b) CTE values of composites as functions of CNF content.

3.5. Flame Retardancy

The CNF xerogels used in this study have a flame self-extinguishing functionality [17]. Thus, the composites were expected to inherit a similar functionality. After being exposed to flame for 3 s, the pristine polymer was rapidly and completely consumed by the flame (Figure 7a and Video S1). By contrast, the CNF composite with 30–42.5% CNF burned slowly, and a black residue remained (Figures 7a and S4a, and Video S2). Interestingly, for the composites containing more than 55 vol% CNF, no flame ignition was observed (Figures 7a and S4a, and Video S3). The non-combusted specimens were further exposed to a flame for 6 s. For the composites with 67.5% and 80 vol% CNF, the ignited flame was self-extinguished (Figure S4b and Video S4).

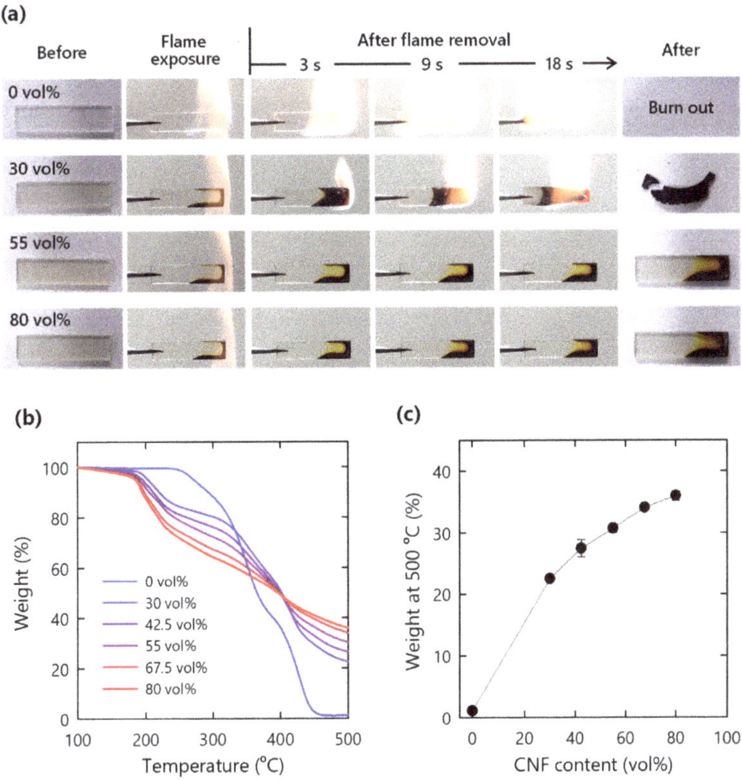

Figure 7. (**a**) Flammability test of CNF composites; (**b**) TG curves of CNF composites under nitrogen conditions; (**c**) residue weight at 500 °C based on TG analysis under nitrogen conditions.

Figure 7b shows the TG curves under nitrogen conditions. The CNF composites started to degrade at approximately 200 °C prior to polymer degradation at 250 °C (Figure 7b). The initial thermal degradation of the composite was derived based on the decarboxylation of TEMPO-oxidized cellulose [34,35]. The non-flammable carbon dioxide emitted from cellulose possibly dilutes the flammable gases [36,37]. The residue weight at 500 °C increased as the CNF content increased (Figure 7c). In contrast, the pristine polymer completely degraded to volatile gas at 450 °C; the CNF composite with 80 vol% CNF retained 35% of its weight at 500 °C. The increase in the residual weight is likely explained by the thermally stable char formation promoted by metal ions on the cellulose fiber surface [35–38]. In addition, aluminum hydroxide structure on the CNF surface dehydrates into aluminum oxide via an endothermic reaction during flame exposure [17]. The formed involatile residues, including char and aluminum oxide, should contribute to the flame retardancy of the CNF composites [36,37]. Meanwhile, the TG curves under air conditions demonstrated similar trends to those under nitrogen conditions (Figure S4c,d). This indicates that oxidation is suppressed, possibly because of barrier function of CNF and CNF char.

4. Conclusions

In this study, thick CNF/polymer composites were prepared via an impregnation method using nanocellulose xerogels. The composite exhibited high optical transmittance over a broad range of CNF content. Analysis of the relationship of the transmittance with thickness suggested that the composite has a homogeneous structure with well-dispersed

CNF fibrils. The fine CNF network efficiently reinforced the polymer matrix, resulting in improvements in modulus, strength, and work of fracture. The network also contributed to the thermal stability of the composite, with a reduction in the CTE value of up to 78%. Additionally, the wood-derived nanofibers endowed the composite with flame retardancy. These unique features highlight the applicability of CNF xerogels as a reinforcing template for producing multifunctional and load-bearing polymer composites.

Supplementary Materials: The following are available online at https://www.mdpi.com/article/10.3390/nano11113032/s1, Figure S1: FTIR spectrum and AFM height image of the CNFs, and appearance of CNF xerogels, Figure S2: Flexural strength, work of fracture, and fractured surfaces of CNF composites, Figure S3: X-ray diffraction patterns of composites, Figure S4: Flammability test for CNF composites and TG data under air conditions, Video S1: Flammability test for pristine polymer, Video S2: Flammability test for CNF composite (30 vol%), Video S3: Flammability test for CNF composite (55 vol%), Video S4: Additional flammability test for CNF composite (80 vol%).

Author Contributions: Conceptualization, T.S., L.A.B. and W.S.; experimental design, investigation, W.S. W.S. mainly wrote the manuscript with contributions from S.F., L.A.B. and T.S. All authors have read and agreed to the published version of the manuscript.

Funding: This research was funded by JST-Mirai R & D Program (JPMJMI17ED), JSPS Grant-in-Aids for Scientific Research (A) (21H04733), and JSPS Grant-in-Aids for Young Scientists (19J22463, 20K15567). LB is supported by the European Research Council (ERC), grant agreement No. 742733.

Data Availability Statement: The data that support the findings of this study are available from the corresponding authors upon reasonable request.

Acknowledgments: TEMPO-oxidized pulp was kindly provided by DKS Co. Ltd.

Conflicts of Interest: The authors declare no conflict of interest.

References

1. Harito, C.; Bavykin, D.V.; Yuliarto, B.; Dipojono, H.K.; Walsh, F.C. Polymer Nanocomposites Having a High Filler Content: Synthesis, Structures, Properties, and Applications. *Nanoscale* **2019**, *11*, 4653–4682. [CrossRef]
2. Moon, R.J.; Martini, A.; Nairn, J.; Simonsen, J.; Youngblood, J. Cellulose Nanomaterials Review: Structure, Properties and Nanocomposites. *Chem. Soc. Rev.* **2011**, *40*, 3941. [CrossRef]
3. Clarkson, C.M.; El Awad Azrak, S.M.; Forti, E.S.; Schueneman, G.T.; Moon, R.J.; Youngblood, J.P. Recent Developments in Cellulose Nanomaterial Composites. *Adv. Mater.* **2020**, *2000718*, 14–17. [CrossRef] [PubMed]
4. Oksman, K.; Aitomäki, Y.; Mathew, A.P.; Siqueira, G.; Zhou, Q.; Butylina, S.; Tanpichai, S.; Zhou, X.; Hooshmand, S. Review of the Recent Developments in Cellulose Nanocomposite Processing. *Compos. Part A Appl. Sci. Manuf.* **2016**, *83*, 2–18. [CrossRef]
5. Fujisawa, S.; Togawa, E.; Kimura, S. Large Specific Surface Area and Rigid Network of Nanocellulose Govern the Thermal Stability of Polymers: Mechanisms of Enhanced Thermomechanical Properties for Nanocellulose/PMMA Nanocomposite. *Mater. Today Commun.* **2018**, *16*, 105–110. [CrossRef]
6. Forti, E.S.; Moon, R.J.; Schueneman, G.T.; Youngblood, J.P. Transparent Tempo Oxidized Cellulose Nanofibril (TOCNF) Composites with Increased Toughness and Thickness by Lamination. *Cellulose* **2020**, *27*, 4389–4405. [CrossRef]
7. Hietala, M.; Mathew, A.P.; Oksman, K. Bionanocomposites of Thermoplastic Starch and Cellulose Nanofibers Manufactured Using Twin-Screw Extrusion. *Eur. Polym. J.* **2013**, *49*, 950–956. [CrossRef]
8. Ansari, F.; Berglund, L.A. Toward Semistructural Cellulose Nanocomposites: The Need for Scalable Processing and Interface Tailoring. *Biomacromolecules* **2018**, *19*, 2341–2350. [CrossRef] [PubMed]
9. Dufresne, A. Cellulose Nanomaterial Reinforced Polymer Nanocomposites. *Curr. Opin. Colloid Interface Sci.* **2017**, *29*, 1–8. [CrossRef]
10. Capadona, J.R.; Van Den Berg, O.; Capadona, L.A.; Schroeter, M.; Rowan, S.J.; Tyler, D.J.; Weder, C. A Versatile Approach for the Processing of Polymer Nanocomposites with Self-Assembled Nanofibre Templates. *Nat. Nanotechnol.* **2007**, *2*, 765–769. [CrossRef]
11. Yano, H.; Sugiyama, J.; Nakagaito, A.N.; Nogi, M.; Matsuura, T.; Hikita, M.; Handa, K. Optically Transparent Composites Reinforced with Networks of Bacterial Nanofibers. *Adv. Mater.* **2005**, *17*, 153–155. [CrossRef]
12. Ansari, F.; Skrifvars, M.; Berglund, L. Nanostructured Biocomposites Based on Unsaturated Polyester Resin and a Cellulose Nanofiber Network. *Compos. Sci. Technol.* **2015**, *117*, 298–306. [CrossRef]
13. Nakagaito, A.N.; Yano, H. Novel High-Strength Biocomposites Based on Microfibrillated Cellulose Having Nano-Order-Unit Web-like Network Structure. *Appl. Phys. A Mater. Sci. Process.* **2005**, *80*, 155–159. [CrossRef]
14. Iwamoto, S.; Nakagaito, A.N.; Yano, H.; Nogi, M. Optically Transparent Composites Reinforced with Plant Fiber-Based Nanofibers. *Appl. Phys. A Mater. Sci. Process.* **2005**, *81*, 1109–1112. [CrossRef]

15. Li, Y.; Fu, Q.; Yu, S.; Yan, M.; Berglund, L. Optically Transparent Wood from a Nanoporous Cellulosic Template: Combining Functional and Structural Performance. *Biomacromolecules* **2016**, *17*, 1358–1364. [CrossRef]
16. Yamasaki, S.; Sakuma, W.; Yasui, H.; Daicho, K.; Saito, T.; Fujisawa, S.; Isogai, A.; Kanamori, K. Nanocellulose Xerogels With High Porosities and Large Specific Surface Areas. *Front. Chem.* **2019**, *7*, 1–8. [CrossRef] [PubMed]
17. Sakuma, W.; Yamasaki, S.; Fujisawa, S.; Kodama, T.; Shiomi, J.; Kanamori, K.; Saito, T. Mechanically Strong, Scalable, Mesoporous Xerogels of Nanocellulose Featuring Light Permeability, Thermal Insulation, and Flame Self-Extinction. *ACS Nano* **2021**, *15*, 1436–1444. [CrossRef]
18. Saito, T.; Nishiyama, Y.; Putaux, J.L.; Vignon, M.; Isogai, A. Homogeneous Suspensions of Individualized Microfibrils from TEMPO-Catalyzed Oxidation of Native Cellulose. *Biomacromolecules* **2006**, *7*, 1687–1691. [CrossRef] [PubMed]
19. ASTM D1003-21. *Standard Test Method for Haze and Luminous Transmittance of Transparent Plastics*; ASTM International: West Conshohocken, PA, USA, 2021.
20. Faure, B.; Salazar-Alvarez, G.; Ahniyaz, A.; Villaluenga, I.; Berriozabal, G.; De Miguel, Y.R.; Bergström, L. Dispersion and Surface Functionalization of Oxide Nanoparticles for Transparent Photocatalytic and UV-Protecting Coatings and Sunscreens. *Sci. Technol. Adv. Mater.* **2013**, *14*. [CrossRef]
21. Zhao, M.; Ansari, F.; Takeuchi, M.; Shimizu, M.; Saito, T.; Berglund, L.A.; Isogai, A. Nematic Structuring of Transparent and Multifunctional Nanocellulose Papers. *Nanoscale Horiz.* **2018**, *3*, 28–34. [CrossRef]
22. Saito, T.; Uematsu, T.; Kimura, S.; Enomae, T.; Isogai, A. Self-Aligned Integration of Native Cellulose Nanofibrils towards Producing Diverse Bulk Materials. *Soft Matter* **2011**, *7*, 8804–8809. [CrossRef]
23. Parlak, O.; Demir, M.M. Anomalous Transmittance of Polystyrene–Ceria Nanocomposites at High Particle Loadings. *J. Mater. Chem. C* **2013**, *1*, 290–298. [CrossRef]
24. Chen, H.; Baitenov, A.; Li, Y.; Vasileva, E.; Popov, S.; Sychugov, I.; Yan, M.; Berglund, L. Thickness Dependence of Optical Transmittance of Transparent Wood: Chemical Modification Effects. *ACS Appl. Mater. Interfaces* **2019**, *11*, 35451–35457. [CrossRef] [PubMed]
25. Sharma, B.; Gatóo, A.; Bock, M.; Ramage, M. Engineered Bamboo for Structural Applications. *Constr. Build. Mater.* **2015**, *81*, 66–73. [CrossRef]
26. Lee, K.Y.; Aitomäki, Y.; Berglund, L.A.; Oksman, K.; Bismarck, A. On the Use of Nanocellulose as Reinforcement in Polymer Matrix Composites. *Compos. Sci. Technol.* **2014**, *105*, 15–27. [CrossRef]
27. Shir Mohammadi, M.; Hammerquist, C.; Simonsen, J.; Nairn, J.A. The Fracture Toughness of Polymer Cellulose Nanocomposites Using the Essential Work of Fracture Method. *J. Mater. Sci.* **2016**, *51*, 8916–8927. [CrossRef]
28. Gao, H.; Qiang, T. Fracture Surface Morphology and Impact Strength of Cellulose/PLA Composites. *Materials* **2017**, *10*, 624. [CrossRef]
29. Nogi, M.; Ifuku, S.; Abe, K.; Handa, K.; Nakagaito, A.N.; Yano, H. Fiber-Content Dependency of the Optical Transparency and Thermal Expansion of Bacterial Nanofiber Reinforced Composites. *Appl. Phys. Lett.* **2006**, *88*, 1–4. [CrossRef]
30. Okahisa, Y.; Yoshida, A.; Miyaguchi, S.; Yano, H. Optically Transparent Wood–Cellulose Nanocomposite as a Base Substrate for Flexible Organic Light-Emitting Diode Displays. *Compos. Sci. Technol.* **2009**, *69*, 1958–1961. [CrossRef]
31. Nakagaito, A.N.; Yano, H. The Effect of Fiber Content on the Mechanical and Thermal Expansion Properties of Biocomposites Based on Microfibrillated Cellulose. *Cellulose* **2008**, *15*, 555–559. [CrossRef]
32. Diaz, J.A.; Wu, X.; Martini, A.; Youngblood, J.P.; Moon, R.J. Thermal Expansion of Self-Organized and Shear-Oriented Cellulose Nanocrystal Films. *Biomacromolecules* **2013**, *14*, 2900–2908. [CrossRef]
33. Hirano, T.; Mitsuzawa, K.; Ishioka, S.; Daicho, K.; Soeta, H.; Zhao, M.; Takeda, M.; Takai, Y.; Fujisawa, S.; Saito, T. Anisotropic Thermal Expansion of Transparent Cellulose Nanopapers. *Front. Chem.* **2020**, *8*, 6–11. [CrossRef] [PubMed]
34. Fukuzumi, H.; Saito, T.; Okita, Y.; Isogai, A. Thermal Stabilization of TEMPO-Oxidized Cellulose. *Polym. Degrad. Stab.* **2010**, *95*, 1502–1508. [CrossRef]
35. Lichtenstein, K.; Lavoine, N. Toward a Deeper Understanding of the Thermal Degradation Mechanism of Nanocellulose. *Polym. Degrad. Stab.* **2017**, *146*, 53–60. [CrossRef]
36. Geng, C.; Zhao, Z.; Xue, Z.; Xu, P.; Xia, Y. Preparation of Ion-Exchanged TEMPO-Oxidized Celluloses as Flame Retardant Products. *Molecules* **2019**, *24*, 1947. [CrossRef] [PubMed]
37. Guo, X.; Wang, Y.; Ren, Y.; Liu, X. Fabrication of Flame Retardant Lyocell Fibers Based on Carboxymethylation and Aluminum Ion Chelation. *Cellulose* **2021**, *28*, 6679–6698. [CrossRef]
38. Shi, R.; Tan, L.; Zong, L.; Ji, Q.; Li, X.; Zhang, K.; Cheng, L.; Xia, Y. Influence of Na$^+$ and Ca^{2+} on Flame Retardancy, Thermal Degradation, and Pyrolysis Behavior of Cellulose Fibers. *Carbohydr. Polym.* **2017**, *157*, 1594–1603. [CrossRef]

 *nanomaterials*

MDPI

Article

Facile Preparation of Mechanically Robust and Functional Silica/Cellulose Nanofiber Gels Reinforced with Soluble Polysaccharides

Marco Beaumont [1,*], Elisabeth Jahn [1], Andreas Mautner [2], Stefan Veigel [3], Stefan Böhmdorfer [1], Antje Potthast [1], Wolfgang Gindl-Altmutter [3] and Thomas Rosenau [2,4,*]

1. Department of Chemistry, Institute of Chemistry of Renewable Resources, University of Natural Resources and Life Sciences Vienna, Konrad-Lorenz-Straße 24, 3430 Tulln, Austria; elisabeth.jahn@students.boku.ac.at (E.J.); stefan.boehmdorfer@boku.ac.at (S.B.); antje.potthast@boku.ac.at (A.P.)
2. Faculty of Chemistry, Institute of Materials Chemistry and Research, Polymer and Composite Engineering (PaCE) Group, University of Vienna, Währinger Street 42, 1090 Vienna, Austria; andreas.mautner@univie.ac.at
3. Department of Material Sciences and Process Engineering, Institute of Wood Technology and Renewable Materials, University of Natural Resources and Life Sciences Vienna, Konrad-Lorenz-Straße 24, 3430 Tulln, Austria; stefan.veigel@boku.ac.at (S.V.); wolfgang.gindl-altmutter@boku.ac.at (W.G.-A.)
4. Johan Gadolin Process Chemistry Centre, Åbo Akademi University, Porthansgatan 3, FI-20500 Turku, Finland
* Correspondence: marcobeaumont1@gmail.com (M.B.); thomas.rosenau@boku.ac.at (T.R.)

Citation: Beaumont, M.; Jahn, E.; Mautner, A.; Veigel, S.; Böhmdorfer, S.; Potthast, A.; Gindl-Altmutter, W.; Rosenau, T. Facile Preparation of Mechanically Robust and Functional Silica/Cellulose Nanofiber Gels Reinforced with Soluble Polysaccharides. *Nanomaterials* **2022**, *12*, 895. https://doi.org/10.3390/nano12060895

Academic Editor: Fabien Grasset

Received: 1 February 2022
Accepted: 3 March 2022
Published: 8 March 2022

Abstract: Nanoporous silica gels feature extremely large specific surface areas and high porosities and are ideal candidates for adsorption-related processes, although they are commonly rather fragile. To overcome this obstacle, we developed a novel, completely solvent-free process to prepare mechanically robust CNF-reinforced silica nanocomposites via the incorporation of methylcellulose and starch. Significantly, the addition of starch was very promising and substantially increased the compressive strength while preserving the specific surface area of the gels. Moreover, different silanes were added to the sol/gel process to introduce in situ functionality to the CNF/silica hydrogels. Thereby, CNF/silica hydrogels bearing carboxyl groups and thiol groups were produced and tested as adsorber materials for heavy metals and dyes. The developed solvent-free sol/gel process yielded shapable 3D CNF/silica hydrogels with high mechanical strength; moreover, the introduction of chemical functionalities further widens the application scope of such materials.

Keywords: nanocellulose; functional nanocomposite; aqueous process; sol–gel; hydrogels; aerogels; freeze-drying; cryogels

1. Introduction

Nowadays, biodegradable and renewable materials play an important role in research and industry. The most abundant biopolymer on Earth, cellulose, has been used by our society for centuries and is indispensable in today's economies. In cellulose-related research, the interest in nanostructured celluloses is steadily rising. This group of celluloses includes cellulose nanocrystals (CNC), bacterial cellulose (BC), and cellulose nanofibrils (CNF) [1,2]. Significantly, CNF and BC, as high aspect ratio nanocelluloses, are frequently used to reinforce composite materials, e.g., 3D silica composites [3,4]. Three-dimensional silica composites based on alkoxysilanes are usually prepared from tetraethyl orthosilicate (TEOS) via sol–gel chemistry in a two-step gelation process: (A) TEOS is hydrolyzed in the presence of an acid catalyst, an organic solvent, and water (usually EtOH/water) [5], yielding a low-viscous precursor consisting of partially or completely hydrolyzed TEOS monomers and oligomers, i.e., the silica sol [6–8]. (B) The gelation of this precursor is then induced under basic conditions, e.g., through the addition of a base catalyst, which causes condensation reactions and the formation of a siloxane polymer network, i.e., the silica gel [5]. The preparation of silica gels via sol–gel chemistry has many advantages: it is a robust and well-investigated method, and functional gels can be easily obtained using special silanes (e.g., 3-aminopropyl triethoxysilane and 3-mercaptopropyl trimethoxysilane) to tune the properties of the end products [9]. Moreover, a variation of the synthesis and drying parameters gives access to a variety of materials, such as fine powders, ceramics, glasses, spherical particles, and nanocomposites gels [5]. Pure silica aerogels feature large specific surface areas (SSAs) ranging from 500 to 1000 m^2/g [10]. However, the application of pure silica gels is limited due to their fragility [11]. Their mechanical properties can be improved through the introduction of polymeric materials [12] or (nano)celluloses [3,4,13,14]. CNF-based and BC-based silica composites were produced through the immersion of porous cellulose aerogels into the silica sol followed by subsequent gelation induced by the addition of ammonia acting as a base catalyst. The main drawbacks of these methods are: (1) The required drying step to produce a cellulose aerogel precursor is time-consuming and limits the processability into arbitrary 3D shapes. (2) Most methods are based on organic solvents, whose use should be avoided or at least reduced for sustainability reasons. (3) Finally, CNF–silica nanocomposites are handled as dry aerogels, and the direct preparation and utilization of CNF–silica hydrogels remains largely untapped.

The focus of this work lies in the fabrication of shapable, mechanically robust CNF–silica hydrogels, using an environmentally benign and solvent-free process. We studied first the influence of the solvent on the properties of the gels and tested methylcellulose (MC) and starch as an additive to increase the compressive strength of CNF–silica gels. Finally, we demonstrate the versatility of our approach through the in situ introduction of functional silanes bearing thiol or carboxylate groups and examined the functionalized gels for utilization as adsorbers for heavy metals and dyes.

2. Materials and Methods

2.1. Materials

Cellulose nanofibers were produced from never-dried bleached birch pulp through disintegration in an M110P microfluidizer (Microfluidics Crop., Newton, MA, USA). The pulp was fibrillated by 12 passes through the microfluidizer. Methylcellulose (product number: M0262, 413 cps, 30.3% methoxyl content) and potato starch (product number: S4251, 25.9 μm granule size, 31% amylose, DP_w of amylopectin subfraction: 35) [15] were purchased from Sigma-Aldrich (Sigma-Aldrich Chemie GmbH, Munich, Germany). *N*-[3-Trimethoxysilyl)propyl] ethylenediamine triacetic acid trisodium salt was purchased from abcr (abcr GmbH, Karlsruhe, Germany) as a 45% aqueous solution. All other used chemicals were, if not otherwise noted, purchased from Sigma-Aldrich with a purity of at least 99%.

2.2. Methods

2.2.1. Preparation of CNF–Silica Hydrogels

Here, 12 mL of 1 wt% CNF dispersion (0.12 g dry cellulose), TEOS (98% purity, 2 mL, 1.9 g, 9.1 mmol), and aqueous HCl (0.16 mL, 0.29 M, 46 μmol HCl) were stirred overnight to hydrolyze TEOS to yield the CNF–silica sol. Then, 0.85 mL of 0.1 mol/L NH_3 (85 μmol) were added and mixed quickly to start the condensation. The solution was filled into syringes to obtain hydrogels with cylindric shape (the shape can be freely adjusted by using different molds or other processing techniques). After complete gelation, the silica nanocomposites were aged in water at 50 °C for at least 10 h to stiffen the silica gel network. The gels were stored in DI water in the fridge.

2.2.2. Preparation of CNF–MC–Silica Hydrogels

Here, 7 mL of 1.7 wt% CNF dispersion (0.12 g of dry cellulose), 3.5 mL of water, methylcellulose (1.5 mL, 2 wt%, 0.03 g), 2 TEOS (2 mL, 1.9 g, 9.1 mmol), and HCl (0.16 mL, 0.29 M, 46 μmol HCl) were stirred overnight to hydrolyze TEOS. Then, 0.85 mL of 0.1 mol/L

NH$_3$ (85 μmol) was added and mixed quickly to start the condensation. The solution was filled into syringes to obtain hydrogels with cylindrical shapes. Finally, CNF–MC–Silica hydrogels were aged in water at 50 °C for at least 10 h to stiffen the network.

2.2.3. Preparation of CNF–Starch–Silica Hydrogels

TEOS (2 mL; 1.9 g, 9.1 mmol) was hydrolyzed in the presence of 7 mL of 1.7 wt% CNF (0.12 g dry cellulose), HCl (0.16 mL, 0.29 M, 46 μmol HCl), and 4 mL of water. Then, a 3 wt% dispersion of starch was prepared by adding the starch into cold deionized water and heating under agitation to 80 °C to allow gelatinization. Then, 1 mL of this mixture was cooled down to 60 °C at room temperature, and then directly added into the CNF–silica sol. Then, 0.85 mL of 0.1 mol/L NH$_3$ (85 μmol) was added and mixed quickly to start the condensation. The solution was filled into syringes to obtain hydrogels of a cylindrical shape. Finally, CNF–Starch–Silica hydrogels were aged in water at 50 °C for at least 10 h to stiffen the network.

2.2.4. Preparation of Thiol-Functionalized CNF–Starch–Silica Hydrogels

The respective hydrogels were prepared according to the above procedure for the preparation of CNF–silica hydrogels, using the same quantities of reactants. After acidic hydrolysis, (3-mercaptopropyl)trimethoxysilane (MTMS) (95% purity, 0.13 mL, 0.14 g, 0.7 mmol) was added into the prepared CNF–silica sol (before the addition of ammonia). Upon addition of the mercapto silane, the sample was protected from light with aluminum foil. Then, 0.85 mL of 0.1 mol/L NH$_3$ (85 μmol) was added, and the sample was transferred into a mold. After complete gelation, the silica nanocomposites were aged in water at 50 °C for at least 10 h to stiffen the network. The gels were stored in DI water in the fridge.

2.2.5. Preparation of Carboxylate-Functionalized CNF–Starch–Silica Hydrogels

The respective hydrogels were prepared according to the above procedure for the preparation of CNF–silica hydrogels, using the same quantities and procedure. Directly after the addition of ammonia, the functional silane, *N*-[3-trimethoxysilyl)propyl] thylenedi-amine triacetic acid trisodium salt (0.13 mL, 0.16 g, 0.3 mmol), was added to the CNF–silica mixture. The sample was directly transferred into a mold. After complete gelation, the silica nanocomposites were aged in water at 50 °C for at least 10 h to stiffen the network. The gels were stored in DI water in the fridge.

2.2.6. Compression Tests

The mechanical properties of the composites were measured on a universal testing machine Zwick/Roell Z020 (Ulm, Germany). Compression tests were performed in a wet state, with a 500 N load cell. The strain rate was set to 2.4 mm/min and samples were compressed to 30%. The compressive strength was defined as the maximum stress in the performed strain range. Measurements were performed in triplicate and compared with Student's *t*-test (unpaired, $n = 3$).

2.2.7. Solvent Exchange to *Tert*-BuOH and Freeze-Drying

Samples were freeze-dried from the respective *t*BuOH gel to avoid ice-templating effects and to preserve the gels' nanostructure upon the freeze-drying process [16]. The samples were solvent-exchanged first with 1:1 *t*BuOH/water, second with 8:2 *t*BuOH/water, and finally with pure *t*BuOH (each solvent exchange step was conducted with solvent amounts of approx. ten times the respective sample volume and equilibrated for 24 h on a shaker). All samples were freeze-dried in a Christ Beta 1–8 LD Plus freeze-dryer. After the last step of solvent exchange with 100% *t*BuOH, the samples were taken out of the system and frozen at −80 °C in a glass vial. The frozen samples were quickly transferred into the lyophilizer. All samples were freeze-dried for at least 24 h and stored afterward in an airtight container.

2.2.8. Specific Surface Area

The 3D composites for the BET measurement were prepared by cutting the dried samples into small pieces, pre-drying them at 60 °C for at least 24 h, and storing them in an airtight beaker with drying beads to keep the samples dry. The samples were degassed in a FlowPrep 060 (Mircomeritics, Norcross, GA, USA) at 80 °C under N_2 flow for at least 6 h. Afterward, the measurement was performed on a Micromeritics TriStarII instrument.

2.2.9. Porosity of Samples

The porosity of the 3D composites was calculated according to Equation (1).

$$\Phi = \left(1 - \frac{\rho}{w_s * \rho_s + w_c * \rho_c}\right) * 100 \tag{1}$$

where ρ is the bulk density of the sample (measured by gravimetric means from dried samples with defined volume), and the porosity was calculated in percent. ρ_s is the density of silica, at 2.19 g/cm^3 [17], and ρ_c is the density of cellulose, at 1.59 g/cm^3 [18]. w_s and w_c are the mass fraction of silica and cellulose, respectively, calculated using Equation (2).

$$w_s[\%] = \left(\frac{\rho - \rho_0}{\rho}\right) * 100 \tag{2}$$

where ρ_0 stands for the cellulose bulk density in the sample, calculated from native CNF or BC gels cryogels. $\rho_0 = 0.010$ g cm^{-3} in case of CNF and 0.007 g cm^{-3} in case of BC.

2.2.10. Scanning Electron Microscopy (SEM)

Micrographs of freeze-dried samples were measured on a Zeiss Supra 55 VP. Before the measurement, the samples were sputtered with a 10 nm-thick gold layer (Leica EM SCD050, Wetzlar, Germany).

2.2.11. Detection of Thiol Groups

The number of thiol groups in the composite sample obtained upon the addition of MTMS was quantified using Ellman's test according to the instructions of Thermo Scientific [19]. Ellman's test was performed in a reaction buffer containing 0.1 M sodium phosphate solution and 1 mM of ethylenediaminetetraacetic acid at a pH of 8. Ellman's reagent solution was prepared as follows: 4 mg Ellman's reagent (10 μmol), i.e., 5,5′-dithio-bis-(2-nitrobenzoic acid), was dissolved in 1 mL of the reaction buffer. A solution of 2.5 mL of reaction buffer and 50 μL of Ellman's reagent solution were prepared for each sample (including blank and positive control). For the blank sample, 250 μL of reaction buffer was added to the prepared solution. For the positive control, 4 μmol of MTMS and 246 μL of reaction buffer were added. To measure the number of thiols in the hydrogels, 10–40 mg hydrogel of each sample was added to the prepared solutions and the volume was adjusted to 2.8 mL with reaction buffer. The total thiol content was detected by UV-VIS spectroscopy (Perkin Elmer Lambda 35, Waltham, MA, USA) at a wavelength of 412 nm.

2.2.12. Adsorption of Copper(II) Sulfate

A copper(II) sulfate solution with a concentration of 500 ppm was prepared using copper (II) sulfate pentahydrate. Here, 8.5 mL of this solution was added to the hydrogel sample (0.9–1.2 g of hydrogel weight with approx. 6% solid content). The performance of the carboxylate-functionalized CNF–silica composite was compared to native CNF–silica composite and a blank sample. The amount of adsorbed copper(II) was measured by monitoring the residual copper(II) concentration in the solution after 8 days of equilibration, which was determined by UV-VIS spectroscopy at 812 nm. The UV-VIS measurements for this sample were carried out without further dilution.

3. Results and Discussion

All composites were prepared via sol–gel chemistry starting from tetraethyl orthosilicate (TEOS). In combination with nanocellulose, mechanically robust hydrogels were obtained. The first experiments aimed at studying the effect of EtOH/water ratio on the mechanical properties and specific surface area (SSA) of cellulose/silica gels. These studies were conducted with bacterial cellulose as a model material since it naturally provided a stiff, pre-shaped and mechanically robust gel network. For follow-up experiments, BC was replaced with CNF, a highly viscous, shapable suspension, which allows an easy adjustment of the final shapes of the gels (Figure 1).

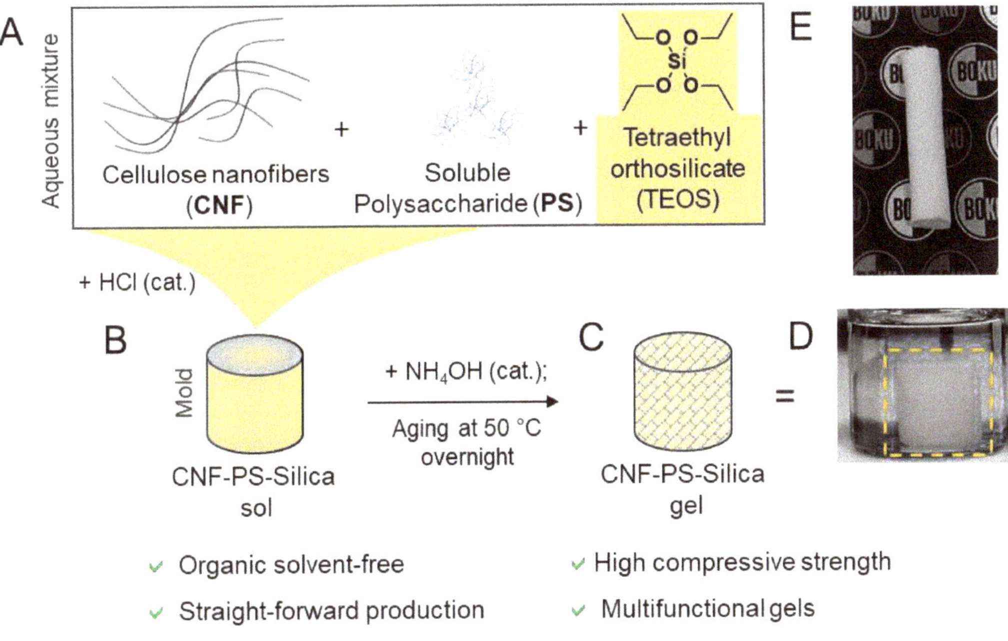

Figure 1. Preparation of silica composite gels reinforced with cellulose nanofibers (CNF) and a soluble polysaccharide (PS), starch, or methylcellulose. (**A**) The individual components are mixed in the presence of catalytic amounts of HCl to catalyze the hydrolysis of tetraethyl orthosilicate (TEOS). (**B**) Subsequently, ammonia is added to increase the pH and trigger the condensation and, thereby, the gelation of the sample. This was followed by the aging of the samples at 50 °C to stiffen the gel network and obtain the final CNF–PS–Silica gel (**C**). The respective hydrogel (**D**) was dried by freeze-drying after solvent exchange to obtain highly porous CNF–PS–silica cryogels (**E**). The shape of the gel can be controlled through molding or 3D printing of the CNF–PS–silica sol.

We compared the gelation in an EtOH/water ratio of 5:1 (*v:v*) with our completely water-based and solvent-free method. The prepared hydrogels were tested with uniaxial compression tests to evaluate their mechanical resistance (Figure S1 and Table S1). In addition, the properties of respective dry gels were compared to study their specific surface area, density, and porosity (Table S2). Cryogels [20] were prepared via freeze-drying after solvent exchange to *t*BuOH [21]. This gives highly porous materials, which are very similar to conventional aerogels prepared via supercritical CO_2 (scCO_2) drying. We selected this approach, since it is more frequently used and easier to perform than drying with scCO_2, and the obtained composite cryogels featured comparable and even slightly higher specific surface area values than supercritical CO_2 dried samples (Table S2). Silica hydrogels—prepared according to our organic, solvent-free method—had significantly higher compressive strength values of 22 kPa in comparison to the 13 kPa of samples obtained using the EtOH/water mixture (Figure S1). The higher compressive strength can be explained by the higher density of the sample produced in the organic, solvent-free process. The main reasons for this effect are most probably the higher rates of the condensation reactions for increasing water content [22], which lead to a larger size of silica particles deposited on the BC structure as shown by scanning electron microscopy (Figure S2). Similar observations have been also made in TEOS model systems, showing that the aggregate/particle size grows with increasing water content during TEOS gelation [23]. The larger particle size in the organic, solvent-free process explains its smaller specific surface area (759 m^2 g^{-1} vs. 897 m^2 g^{-1}) and lower porosity (Table S1). These values are in the range of reported properties of BC silica composites [3]. In comparison to previous approaches [3,4,13], no drying step of nanocellulose is required, and we were able to directly use wet nanocellulose gels. To generate gels with high mechanical strength, we used our organic, solvent-free approach in the subsequent CNF composite preparation.

We replaced BC with CNF to further increase the versatility of our method (Figure 1) and allow the production of silica composite gels of arbitrary shape (Figure 1D,E). In this process, TEOS was hydrolyzed in a CNF suspension in the presence of HCl as an acidic catalyst for the preparation of CNF–silica gels or CNF in combination with soluble polysaccharides (PS; methylcellulose (MC), starch) to obtain CNF–PS–Silica gels. After complete hydrolysis (usually after stirring overnight), the respective sol (Figure 1B) can be transferred into a mold or processed with an extrusion technique, such as 3D printing, for shaping purposes. The addition of ammonia finally triggered the gelation of the silica gels, and the gel network was densified and stiffened (causing gel shrinkage) through curing/aging at 50 °C in water (Figure 1C,D).

The mechanical properties of the gels are compared to a BC–Silica gel (Figure 2 and Table 1). Because of the already-strong gel network of native BC, the prepared BC–Silica gels featured higher compressive strength than the CNF–Silica gel (22 kPa vs. 14 kPa). Since our aim was the preparation of mechanically robust composite gels, we studied the addition of two soluble polysaccharides—starch or MC—to the gel network to strengthen it. The addition of MC slightly raised the compressive strength by approx. 10%, whereas the starch addition almost doubled the compressive strength from 14 kPa (CNF–Silica gel) to 26 kPa (CNF–Starch–Silica gel).

Table 1. Properties of CNF–silica gels and the influence of the addition of starch or methylcellulose (MC). The standard deviation of the average compressive strengths is reported ($n = 3$).

Samples	Density (g cm^{-3})	Specific Surface Area (m^2 g^{-1})	Porosity (%)	Compressive Strength * (kPa)
CNF–Silica	0.061	603	97.1	14 ± 1
CNF–MC–Silica	0.062	740	97.1	15 ± 2
CNF–Starch–Silica	0.065	625	96.9	26 ± 4
CNF	0.010	135	99.0	-

* Measurements were conducted from 0 to 30% strain, and the highest compressive stress value in this range was defined as compressive strength.

Respective cryogels were prepared by freeze-drying, after solvent exchange to *t*BuOH, for analytical purposes. Native CNF cryogels featured a specific surface area of 135 m^2 g^{-1} and a porosity of 99% (Table 1), which is comparable to other values in the literature [24,25]. CNF–Silica gels had a much higher specific surface area of 603 m^2 g^{-1} and density due to the incorporation of the silica network (Figure 3B). MC and starch increased the densities and enlarged the SSA. The increase in SSA was especially evident in the case of CNF–MC–Silica, which increased from 603 m^2/g (CNF–Silica) to 740 m^2/g. As shown in Figure 3, the native CNF network is covered with silica particles in the composite samples. The CNF network structure is visible in the case of CNF–Silica and CNF–MC–Silica cryogels (Figure 3B,C), in which individual fibrils are covered with silica particles. In contrast to that, CNF–Starch–

Silica featured a very different cauliflower-like nanostructure (Figure 3D), which covered nearly completely the fibrillar skeleton of CNF. The drastic change in nanostructure in the case of CNF–Starch–Silica cryogels is in line with the higher density, lower porosity, and increased compressive strength (Table 1).

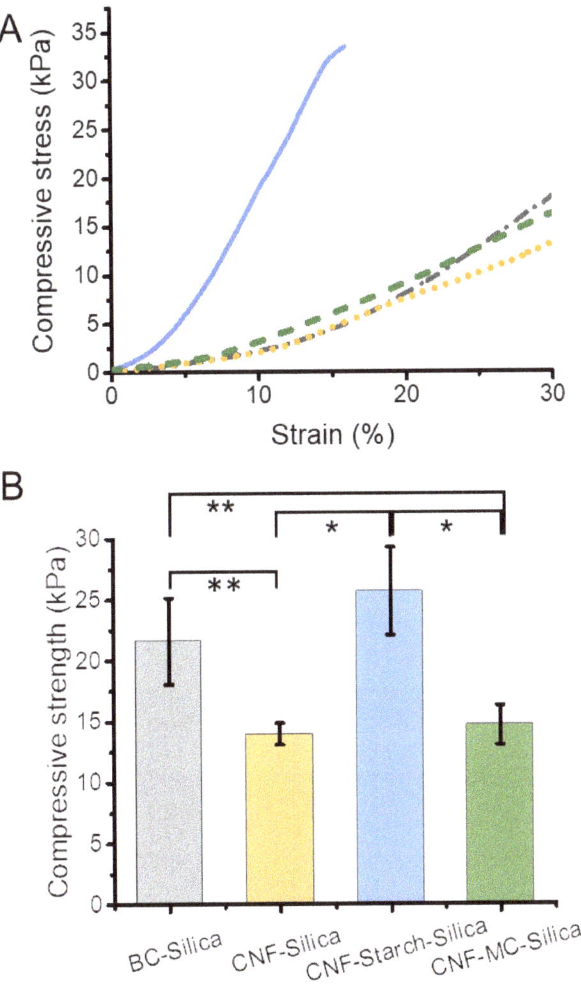

Figure 2. Mechanical properties of the prepared CNF–silica hydrogels and influence of the addition of the soluble polysaccharides—starch or methylcellulose (MC)—on the compression behavior. (**A**) Compression tests of the CNF–Silica hydrogels up to 30% strain (CNF–silica: dotted yellow line, CNF–MC–Silica: green dashed line, and CNF–Starch–Silica: blue solid line) in comparison to bacterial cellulose (BC) silica gel (dash-dotted gray line). (**B**) Comparison of the average compressive strengths and their standard deviation of the samples. Highlighted differences are statistically significant (* $p < 0.04$, ** $p < 0.04$, $n = 3$).

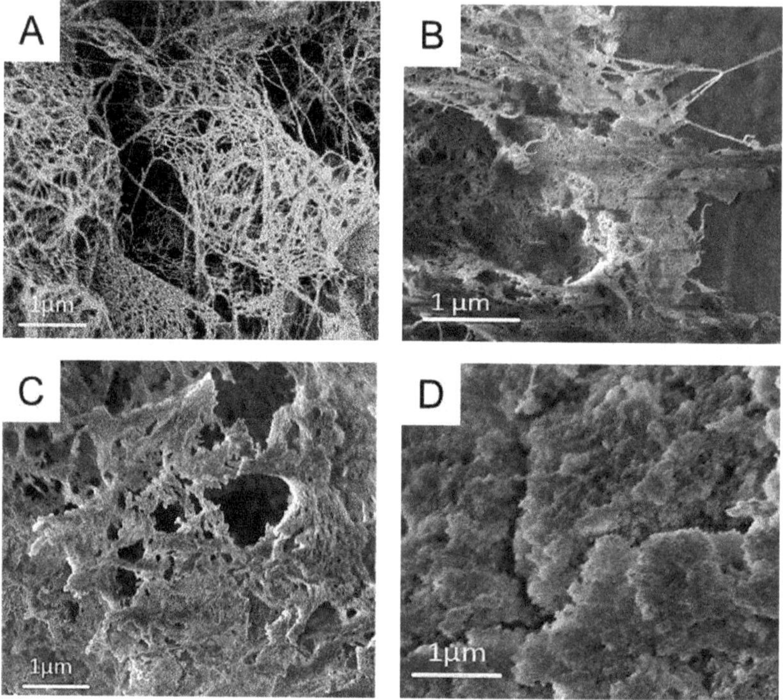

Figure 3. Scanning electron micrographs of CNF (**A**), CNF–Silica (**B**), CNF–Methylcellulose–Silica (**C**), and CNF–Starch–Silica (**D**) cryogels.

As summarized in Figure 1, our proposed method is straightforward, organic, solvent-free, and enables the preparation of moldable CNF–silica gels of high compressive strength. The prepared CNF–silica gels also featured large specific surface areas in the range of 603–740 m^2 g^{-1}.

Besides TEOS, alkoxysilanes are commercially available with a wide range of additional chemical functionalities, and those functional silanes were shown, e.g., to be suitable for the functionalization of pristine CNFs [26–29]. The addition of chemical functional groups onto the surface of the prepared composites would further increase their application range and versatility. We tested the in situ modification of the CNF–Silica composites with alkoxysilanes bearing thiol (Figure 4A1) and carboxylate groups (Figure 4A2).

Functional CNF–Silica gels were prepared with 0.3–0.7 mmol (3–8% functionalization degree based on the total molar amount of TEOS) of the respective alkoxysilane (Figure 4A1,A2), 3-mercaptopropyl trimethoxysilane (MTMS), or *N*-[3-trimethoxysilyl)propyl] ethylenediamine triacetic acid trisodium salt (3CTMS). In the following section, we focus on the results of functional CNF–Silica hydrogels, this protocol can be also applied to the mechanically robust polysaccharide-reinforced composites, CNF–MC–Silica, and CNF–Starch–Silica. Since the sol–gel process of silica gels is largely dependent on the pH, the acidity/basicity of the functional silane must be considered, e.g., an uncontrolled change of pH could prevent gelation or cause uncontrolled gelation and the formation of isolated gelled lumps (this has to be taken into account if the compatibility of our method to other functional silanes is tested). The functional silanes were first added directly into the silica sol after the TEOS hydrolysis step to avoid influences on the hydrolysis and early reaction of TEOS and the functional silane. MTMS did not influence the pH and could be added into the silica sol without any modification. Upon addition of MTMS, the gelation was triggered, as in our standard protocol with ammonia, where we noted that no shrinkage occurred during the aging step. A similar effect was observed in the literature in the case of hydrophobic silanes [30]. Due to the hydrophilicity of the thiol groups and their susceptibility to oxidation, we assume that disulfide crosslinks are partially formed preventing shrinkage. We proved the successful introduction of thiol groups into the composite with Ellman's test, which allows the quantitative detection of thiols [31]. The incorporated amount of thiols was determined to be 1.3 μmol g^{-1} gel. Due to the high reactivity of these groups, they can be post-modified with high efficiency using, e.g., thiol–epoxy [32] or thiol–ene [33] and thiol–Michael [34] click chemistry.

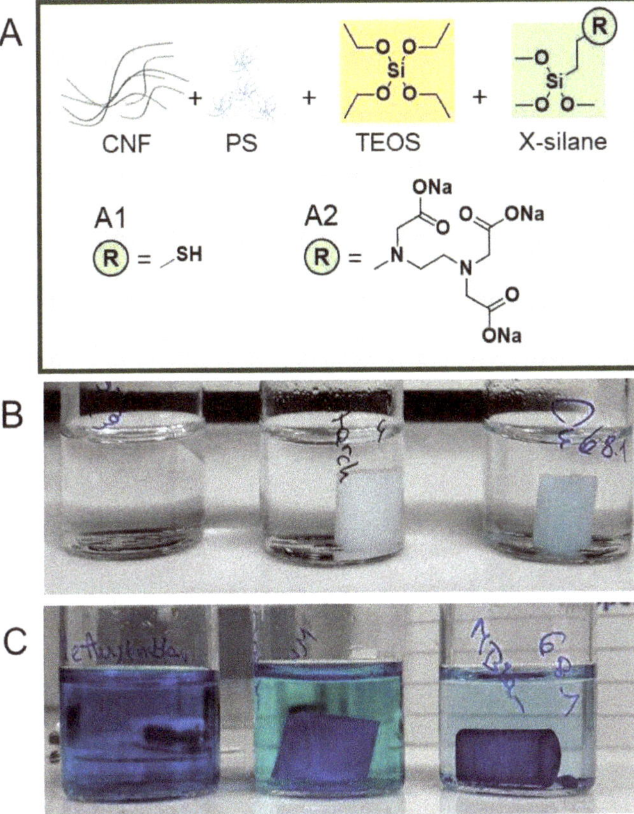

Figure 4. (**A**) Preparation of functional CNF–Silica hydrogels through the addition of either 3-mercaptopropyl trimethoxysilane (**A1**) or *N*-[3-trimethoxysilyl)propyl] ethylenediamine triacetic acid trisodium salt (**A2**) to introduce either thiol or carboxylate structures onto the hydrogel. Carboxylated CNF–Silica gels were tested as adsorbers for Cu(II) ions (**B**) and methylene blue (**C**). Vials in (**B,C**) contain (left to right): solutions before the addition of gels, non-functional CNF–Silica gel, and carboxylated CNF–Silica gel.

Due to the basicity of 3CTMS (carboxylate form), the first tests were unsuccessful and the addition of 3CTMS into CNF–Silica sol neutralized the medium, causing rapid, uncontrolled gelation. To avoid this, we adapted the method and added ammonia directly before 3CTMS. Thereby, the condensation rate was reduced, and we were able to produce stable CNF–Silica gels functionalized with carboxylate groups. We tested these carboxylated silica hydrogels in two different adsorber applications: the adsorption of copper(II) cations and the adsorption of the cationic dye methylene blue. All adsorption experiments were conducted for 8 days and the amount of adsorbed Cu(II) was determined with photometry. The carboxylated CNF–Silica gel adsorbed up 0.58 mg Cu(II)/g of hydrogel, whereas the native CNF–Silica gel was able to bind only 0.04 mg/g (Figure 4B). This demonstrates a higher adsorption capacity of the functional gel and proves as well the successful incorporation of the carboxylate groups. Furthermore, we tested the adsorption of methylene blue (a 20 ppm test solution) as part of our qualitative experiments. Additionally, in this case the dye adsorption capacity of the CNF–silica gel modified with 3CTMS was significantly higher than that of the CNF–silica gel counterpart (Figure 4C).

4. Conclusions

In this contribution, we demonstrated that CNF–Silica hydrogels can be produced straightforwardly with an organic, solvent-free process. In our process, no prior drying step is required, and the CNF can be directly dispersed in the silica sol. Gelation is finally triggered through the addition of ammonia, and the samples can be easily shaped through molding or other processing techniques suitable for CNFs. We further increased the compressive strength of those hydrogels through the addition of soluble polysaccharides. Significantly, the addition of starch was very promising and significantly increased the compressive strength (26 kPa) while preserving the specific surface area of the gels. Depending on the process, the specific surface area of the CNF–silica samples can be tuned from 603 m^2 g^{-1} to 740 m^2 g^{-1}. Finally, we showed that alkoxysilanes bearing thiol and carboxylate groups can be incorporated to introduce functional groups onto the composite gels.

Supplementary Materials: The following supporting information can be downloaded at: https://www.mdpi.com/article/10.3390/nano12060895/s1. Table S1: Influence of different processing conditions with and without EtOH on the properties of the prepared. Table S2: Influence of dying conditions on the properties of bacterial cellulose aerogels. Figure S1: Compressive test of bacterial cellulose silica hydrogel samples prepared in EtOH/water (5:1, *v:v*, light blue line) and pure water (dark blue line). Figure S2: Scanning electron micrographs of A) the sample BC–Silica water/EtOH (1:5, *v:v*) in comparison to B) BC–silica water.

Author Contributions: Conceptualization, M.B. and T.R.; methodology A.P., W.G.-A., M.B., S.B. and T.R.; formal analysis, investigation, data curation, E.J., A.M., S.V. and M.B.; Resources, W.G.-A., A.P. and T.R.; writing and visualization, M.B. and E.J.; supervision, S.B., M.B., A.P. and T.R. All authors have read and agreed to the published version of the manuscript.

Funding: Austrian Science Fund (FWF) (J4356), Austrian Biorefinery Centre Tulln (ABCT).

Data Availability Statement: The data that support the findings of this study are available on request from the corresponding author, M.B.

Acknowledgments: The authors thank the Open Access Funding by the Austrian Science Fund (FWF). For the purpose of open access, the author has applied a CC BY public copyright license to any author accepted manuscript version arising from this submission. The authors thank the financial support from the Austrian Biorefinery Centre Tulln (ABCT). Orlando Rojas is acknowledged for supplying cellulose nanofibers.

Conflicts of Interest: The authors declare no conflict of interest.

References

1. Beaumont, M.; Potthast, A.; Rosenau, T. Cellulose Nanofibrils: From Hydrogels to Aerogels. In *Cellulose Science and Technology*; Rosenau, T., Potthast, A., Hell, J., Eds.; John Wiley & Sons, Inc.: Hoboken, NJ, USA, 2018; pp. 277–339. ISBN 978-1-119-21761-9.
2. Tardy, B.L.; Mattos, B.D.; Otoni, C.G.; Beaumont, M.; Majoinen, J.; Kämäräinen, T.; Rojas, O.J. Deconstruction and Reassembly of Renewable Polymers and Biocolloids into Next Generation Structured Materials. *Chem. Rev.* **2021**, *121*, 14088–14188. [CrossRef] [PubMed]
3. Sai, H.; Xing, L.; Xiang, J.; Cui, L.; Jiao, J.; Zhao, C.; Li, Z.; Li, F. Flexible Aerogels Based on an Interpenetrating Network of Bacterial Cellulose and Silica by a Non-Supercritical Drying Process. *J. Mater. Chem. A* **2013**, *1*, 7963. [CrossRef]
4. Fu, J.; Wang, S.; He, C.; Lu, Z.; Huang, J.; Chen, Z. Facilitated Fabrication of High Strength Silica Aerogels Using Cellulose Nanofibrils as Scaffold. *Carbohydr. Polym.* **2016**, *147*, 89–96. [CrossRef] [PubMed]

5.   Ruiz-Hitzky, E.; Darder, M.; Aranda, P. An Introduction to Bio-nanohybrid Materials. In *Bio-inorganic Hybrid Nanomaterials*; Ruiz-Hitzky, E., Ariga, K., Lvov, Y., Eds.; Wiley-VCH: Weinheim, Germany; John Wiley: Chichester, UK, 2008; ISBN 978-3-527-31718-9.

6.   Brinker, C.J.; Scherer, G.W. *Sol-Gel Science: The Physics and Chemistry of Sol-Gel Processing*; Academic Press: Boston, MA, USA, 1990; ISBN 978-0-12-134970-7.

7.   Brinker, C.J. Hydrolysis and Condensation of Silicates: Effects on Structure. *J. Non-Cryst. Solids* **1988**, *100*, 31–50. [CrossRef]

8.   Schubert, U. Sol-Gel-Chemie: Einfach und doch kompliziert. *Chem. Unserer Zeit* **2018**, *52*, 18–25. [CrossRef]

9.   Alnaief, M.; Smirnova, I. Effect of Surface Functionalization of Silica Aerogel on Their Adsorptive and Release Properties. *J. Non-Cryst. Solids* **2010**, *356*, 1644–1649. [CrossRef]

10. Al-Oweini, R.; El-Rassy, H. Surface Characterization by Nitrogen Adsorption of Silica Aerogels Synthesized from Various Si(OR)$_4$ and R"Si(OR')$_3$ Precursors. *Appl. Surf. Sci.* **2010**, *257*, 276–281. [CrossRef]

11. Wong, J.C.H.; Kaymak, H.; Tingaut, P.; Brunner, S.; Koebel, M.M. Mechanical and Thermal Properties of Nanofibrillated Cellulose Reinforced Silica Aerogel Composites. *Microporous Mesoporous Mater.* **2015**, *217*, 150–158. [CrossRef]

12. Leventis, N. Three-Dimensional Core-Shell Superstructures: Mechanically Strong Aerogels. *Acc. Chem. Res.* **2007**, *40*, 874–884. [CrossRef]

13. Fu, J.; He, C.; Wang, S.; Chen, Y. A Thermally Stable and Hydrophobic Composite Aerogel Made from Cellulose Nanofibril Aerogel Impregnated with Silica Particles. *J. Mater. Sci.* **2018**, *53*, 7072–7082. [CrossRef]

14. Litschauer, M.; Neouze, M.-A.; Haimer, E.; Henniges, U.; Potthast, A.; Rosenau, T.; Liebner, F. Silica Modified Cellulosic Aerogels. *Cellulose* **2011**, *18*, 143–149. [CrossRef]

15. Tester, R.F.; Debon, S.J.; Davies, H.V.; Gidley, M. Effect of Temperature on the Synthesis, Composition and Physical Properties of Potato Starch. *J. Sci. Food Agric.* **1999**, *79*, 2045–2051. [CrossRef]

16. Beaumont, M.; Rennhofer, H.; Opietnik, M.; Lichtenegger, H.C.; Potthast, A.; Rosenau, T. Nanostructured Cellulose II Gel Consisting of Spherical Particles. *ACS Sustain. Chem. Eng.* **2016**, *4*, 4424–4432. [CrossRef]

17. Rumble, J.R.; Lide, D.R.; Bruno, T.J. *CRC Handbook of Chemistry and Physics: A Ready-Reference Book of Chemical and Physical Data*, 99th ed.; CRC Press: Boca Raton, FL, USA, 2018; ISBN 978-1-138-56163-2.

18. Sai, H.; Fu, R.; Xing, L.; Xiang, J.; Li, Z.; Li, F.; Zhang, T. Surface Modification of Bacterial Cellulose Aerogels' Web-like Skeleton for Oil/Water Separation. *ACS Appl. Mater. Interfaces* **2015**, *7*, 7373–7381. [CrossRef]

19. Thermo Scientific. Instructions Ellman's Reagent. Available online: https://assets.fishersci.com/TFS-Assets/LSG/manuals/MAN0011216_Ellmans_Reag_UG.pdf (accessed on 1 March 2022).

20. Buchtová, N.; Budtova, T. Cellulose Aero-, Cryo- and Xerogels: Towards Understanding of Morphology Control. *Cellulose* **2016**, *23*, 2585–2595. [CrossRef]

21. Beaumont, M.; Kondor, A.; Plappert, S.; Mitterer, C.; Opietnik, M.; Potthast, A.; Rosenau, T. Surface Properties and Porosity of Highly Porous, Nanostructured Cellulose II Particles. *Cellulose* **2017**, *24*, 435–440. [CrossRef]

22. Jang, K.W.; Choi, S.H.; Pyun, S.I.; John, M.S. The Effects of the Water Content, Acidity, Temperature and Alcohol Content on the Acidic Sol-Gel Polymerization of Tetraethoxysilane (TEOS) with Monte Carlo Simulation. *Mol. Simul.* **2001**, *27*, 1–16. [CrossRef]

23. Strawbridge, I.; Craievich, A.F.; James, P.F. The Effect of the H$_2$O/TEOS Ratio on the Structure of Gels Derived by the Acid Catalysed Hydrolysis of Tetraethoxysilane. *J. Non-Cryst. Solids* **1985**, *72*, 139–157. [CrossRef]

24. Nechyporchuk, O.; Belgacem, M.N.; Bras, J. Production of Cellulose Nanofibrils: A Review of Recent Advances. *Ind. Crops Prod.* **2016**, *93*, 2–25. [CrossRef]

25. Sehaqui, H.; Zhou, Q.; Berglund, L.A. High-Porosity Aerogels of High Specific Surface Area Prepared from Nanofibrillated Cellulose (NFC). *Compos. Sci. Technol.* **2011**, *71*, 1593–1599. [CrossRef]

26. Hettegger, H.; Beaumont, M.; Potthast, A.; Rosenau, T. Aqueous Modification of Nano- and Microfibrillar Cellulose with a Click Synthon. *ChemSusChem* **2016**, *9*, 75–79. [CrossRef] [PubMed]

27. Beaumont, M.; Bacher, M.; Opietnik, M.; Gindl-Altmutter, W.; Potthast, A.; Rosenau, T. A General Aqueous Silanization Protocol to Introduce Vinyl, Mercapto or Azido Functionalities onto Cellulose Fibers and Nanocelluloses. *Molecules* **2018**, *23*, 1427. [CrossRef] [PubMed]

28. Hettegger, H.; Sumerskii, I.; Sortino, S.; Potthast, A.; Rosenau, T. Silane Meets Click Chemistry: Towards the Functionalization of Wet Bacterial Cellulose Sheets. *ChemSusChem* **2015**, *8*, 680–687. [CrossRef] [PubMed]

29. Cunha, A.G.; Lundahl, M.; Ansari, M.F.; Johansson, L.-S.; Campbell, J.M.; Rojas, O.J. Surface Structuring and Water Interactions of Nanocellulose Filaments Modified with Organosilanes toward Wearable Materials. *ACS Appl. Nano Mater.* **2018**, *1*, 5279–5288. [CrossRef]

30. Moro, S.; Parneix, C.; Cabane, B.; Sanson, N.; d'Espinose de Lacaillerie, J.-B. Hydrophobization of Silica Nanoparticles in Water: Nanostructure and Response to Drying Stress. *Langmuir* **2017**, *33*, 4709–4719. [CrossRef]

31. Bhat, R.; Grover, G. *Ellman's Assay for in-Solution Quantification of Sulfhydryl Groups*; Contraline Inc.: Charlottesville, VA, USA, 2022.

32. Huynh, C.T.; Liu, F.; Cheng, Y.; Coughlin, K.A.; Alsberg, E. Thiol-Epoxy "Click" Chemistry to Engineer Cytocompatible PEG-Based Hydrogel for SiRNA-Mediated Osteogenesis of HMSCs. *ACS Appl. Mater. Interfaces* **2018**, *10*, 25936–25942. [CrossRef]

33. Hoyle, C.E.; Bowman, C.N. Thiol-Ene Click Chemistry. *Angew. Chem. Int. Ed.* **2010**, *49*, 1540–1573. [CrossRef]
34. Nair, D.P.; Podgórski, M.; Chatani, S.; Gong, T.; Xi, W.; Fenoli, C.R.; Bowman, C.N. The Thiol-Michael Addition Click Reaction: A Powerful and Widely Used Tool in Materials Chemistry. *Chem. Mater.* **2014**, *26*, 724–744. [CrossRef]

Article

Polydopamine Doping and Pyrolysis of Cellulose Nanofiber Paper for Fabrication of Three-Dimensional Nanocarbon with Improved Yield and Capacitive Performances

Luting Zhu *, Kojiro Uetani, Masaya Nogi and Hirotaka Koga *

SANKEN (The Institute of Scientific and Industrial Research), Osaka University, 8-1 Mihogaoka,
Ibaraki 567-0047, Osaka, Japan; uetani@eco.sanken.osaka-u.ac.jp (K.U.); nogi@eco.sanken.osaka-u.ac.jp (M.N.)
* Correspondence: sharollzhu@eco.sanken.osaka-u.ac.jp (L.Z.); hkoga@eco.sanken.osaka-u.ac.jp (H.K.);
  Tel.: +81-6-6879-8442 (L.Z. & H.K.)

Abstract: Biomass-derived three-dimensional (3D) porous nanocarbons have attracted much attention due to their high surface area, permeability, electrical conductivity, and renewability, which are beneficial for various electronic applications, including energy storage. Cellulose, the most abundant and renewable carbohydrate polymer on earth, is a promising precursor to fabricate 3D porous nanocarbons by pyrolysis. However, the pyrolysis of cellulosic materials inevitably causes drastic carbon loss and volume shrinkage. Thus, polydopamine doping prior to the pyrolysis of cellulose nanofiber paper is proposed to fabricate the 3D porous nanocarbons with improved yield and volume retention. Our results show that a small amount of polydopamine (4.3 wt%) improves carbon yield and volume retention after pyrolysis at 700 °C from 16.8 to 26.4% and 15.0 to 19.6%, respectively. The pyrolyzed polydopamine-doped cellulose nanofiber paper has a larger specific surface area and electrical conductivity than cellulose nanofiber paper that without polydopamine. Owing to these features, it also affords a good specific capacitance up to 200 F g^{-1} as a supercapacitor electrode, which is higher than the recently reported cellulose-derived nanocarbons. This method provides a pathway for the effective fabrication of high-performance cellulose-derived 3D porous nanocarbons.

Keywords: polydopamine doping; cellulose nanofiber; pyrolysis; 3D porous nanocarbon; supercapacitor

Citation: Zhu, L.; Uetani, K.; Nogi, M.; Koga, H. Polydopamine Doping and Pyrolysis of Cellulose Nanofiber Paper for Fabrication of Three-Dimensional Nanocarbon with Improved Yield and Capacitive Performances. *Nanomaterials* **2021**, *11*, 3249. https://doi.org/10.3390/nano11123249

Academic Editors: Wei Zhang and Vijay Kumar Thakur

Received: 2 November 2021
Accepted: 26 November 2021
Published: 30 November 2021

Publisher's Note: MDPI stays neutral with regard to jurisdictional claims in published maps and institutional affiliations.

1. Introduction

There has been rapid progress in the fabrication and design of three-dimensional (3D) porous nanocarbon materials because they provide several advantages such as high surface areas, short diffusion spaces for fast reaction kinetics, and efficient electron pathways [1,2]. As a result, 3D porous nanocarbons have been actively investigated for energy storage applications [3–5], including supercapacitors [6,7] and batteries [8–10].

The majority of carbon-based materials have been conventionally fabricated using petroleum-based precursors [11]. For example, carbon nanofibers have been produced from polyacrylonitrile, pitches, and phenolic resins [12], while commercial carbon fibers, including carbon nanofibers, are produced from petroleum-based precursors only [13], more than 96% of commercial carbon fibers are made from polyacrylonitrile [13,14]. From the viewpoint of sustainable development, there has been an increased demand to replace these non-renewable petroleum-based precursors with abundant and renewable precursors [15,16]. Therefore, renewable biomass-derived 3D porous nanocarbons have been actively developed [17–20].

Cellulose, which is directly produced from natural plants, is known as the most abundant and renewable carbohydrate polymer on earth [21]. Cellulose intrinsically exists in the plant cell wall [22], and it can be extracted in the form of nanofibers by physical and/or chemical methods [23–25]. Because plant-derived cellulose nanofibers can be fabricated into paper with 3D porous nanostructures by the solvent exchange and papermaking processes [26,27], cellulose nanofiber paper may be a promising precursor to the fabrication of 3D porous nanocarbons by pyrolysis. Cellulose possesses the theoretical carbon content of 44.4 wt%, which refers to the residual six carbon atoms per anhydroglucose unit in the cellulose molecular structure [28]. However, the pyrolysis of cellulose is inevitably accompanied by combustion reactions, forming volatile low-molecular-weight carbon-containing substances such as CO, CO_2, alcohols, and ketones and resulting in low yield and drastic volume shrinkage [28,29]. Thus, it is a challenge to fabricate cellulose-derived 3D porous nanocarbons with improved yield and volume retention.

Thus, in this study, polydopamine doping prior to the pyrolysis of the cellulose nanofiber paper is proposed. The small amount of dopamine, which is a well-known biomolecule that is present in various animals [30] is polymerized in situ in the cellulose nanofiber suspension, and it is then fabricated into paper with 3D porous nanostructures. The resulting polydopamine-doped cellulose nanofiber paper is pyrolyzed to prepare the 3D porous nanocarbon with improved yield and volume retention compared to that pyrolyzed without polydopamine doping. Furthermore, polydopamine doping also yields cellulose nanofiber papers with enhanced specific surface areas and electrical conductivity, thereby providing high capacitive performances for energy storage applications.

2. Materials and Methods

2.1. Materials

Cellulose nanofiber water suspension (BiNFi-s cellulose, raw material: softwood bleached kraft pulp) was obtained from Sugino Machine Ltd., Toyama, Japan. Dopamine hydrochloride, *tert*-butyl alcohol, tris(hydroxymethyl)aminomethane, potassium hydroxide, and 0.5 M hydrochloric acid solution were obtained from Nacalai Tesque, Inc., Kyoto, Japan. All chemicals were of analytical reagent grade and were used without further purification.

2.2. In Situ Polymerization of Dopamine in Cellulose Nanofiber Suspension

In situ polymerization of dopamine in the cellulose nanofiber suspension was performed according to a previous report [31]. Briefly, 0.24 g tris(hydroxymethyl)aminomethane was first added to cellulose nanofiber water suspension (0.2 wt%, 200 mL), and the pH was adjusted to 8.0 by means of a 0.5 M hydrochloric acid solution. Then, different amounts of dopamine hydrochloride (0.025 g, 0.05 g, and 0.1 g) were mixed with the suspension, and the suspension was then stirred for 1 day at approximately 25 °C. The suspension turned black in color during the dopamine polymerization [32,33].

Neat polydopamine was also prepared by adding 0.24 g tris(hydroxymethyl)aminomethane to 200 mL distilled water and by adjusting the pH to 8.0 with 0.5 M hydrochloric acid followed by adding 0.1 g dopamine hydrochloride and stirring the mixture for 1 day at approximately 25 °C. Subsequently, the suspension was centrifugated at 10,000 rpm for 15 min (Model 7000, Kubota Corporation Co., Ltd., Tokyo, Japan). The precipitate was washed with distilled water, and it was then centrifuged in the same conditions; this washing treatment was repeated three times. Finally, the polydopamine was dispersed in distilled water and frozen overnight in a refrigerator (SJ-23T, Sharp Corp., Osaka, Japan) and freeze dried (EYELA FDU-2200, Tokyo Rikakikai Co., Ltd., Tokyo, Japan).

2.3. Preparation of Polydopamine-Doped Cellulose Nanofiber Papers

The aqueous mixture of polydopamine and cellulose nanofibers was dewatered by suction filtration (KST-47, Advantec Toyo Kaisha, Ltd., Tokyo, Japan) through a commercial membrane filter (H020A090C, hydrophilic polytetrafluoroethylene membrane with pore size of 0.2 μm, Advantec Toyo Kaisha, Ltd., Tokyo, Japan). Then, 200 mL distilled water was gently added onto the wet sheet, and it was then vacuum filtrated to remove excess reagents. To form the 3D porous nanostructures [27,34], 200 mL of *tert*-butyl alcohol was further added, followed by vacuum filtration. The obtained wet paper was then peeled off from the filter, stored at approximately −18 °C for over 2 h in a refrigerator, and freeze-dried to prepare the polydopamine-doped cellulose nanofiber papers.

2.4. Pyrolysis of Polydopamine-Doped Cellulose Nanofiber Papers

The polydopamine-doped cellulose nanofiber papers were cut into square-shaped samples with the size of 1.5×1.5 cm^2, followed by pyrolysis at 700 °C for 1 h under N$_2$ gas using a desktop gas convertible vacuum furnace (KDF75, Denken-Highdental Co., Ltd., Kyoto, Japan). The heating and cooling rates were set at 2 °C min^{-1}. The N$_2$ flux was set at 0.5 L min^{-1}. The weight and volume of the original and polydopamine-doped cellulose nanofiber papers were measured before and after pyrolysis. To calculate the weight and volume retention after pyrolysis, five pieces of the paper samples were prepared for each polydopamine content. The polydopamine particles were also pyrolyzed under the same conditions. To calculate the weight retention after pyrolysis, three batches of the polydopamine particles were prepared.

2.5. Electrochemical Tests as Supercapacitor Electrodes

The electrochemical tests as a supercapacitor electrode were performed by a three-electrode system using 6 M KOH aqueous electrolyte (ModuLab XM, Solartron Analytical-AMETEK Advanced Measurement Technology Inc., Berkshire, UK). Approximately 2–5 mg pyrolyzed polydopamine-doped cellulose nanofiber paper was directly evaluated as a working electrode, which was covered by nickel foam for electrical signal collection. A Hg/HgO electrode was used as the reference electrode. A platinum wire was used as the counter electrode. Galvanostatic charge/discharge tests and the cyclic voltamme-try (CV) were operated at the current densities of 0.5–20 A g^{-1} and the scan rate of 10 mV s^{-1}, respectively. Electrochemical impedance spectroscopy (EIS) was measured in the frequency range from 0.1 to 100 kHz, with an amplitude of 5 mV. A single-electrode gravimetric capacitance (C, F g^{-1}) was estimated by the following formula, according to charge/discharge curves:

$$C = IDt/(m\Delta V) \tag{1}$$

where I and Dt are the charge/discharge current (A) and the discharge time (s), respectively, m is the weight (g) of the pyrolyzed polydopamine-doped cellulose nanofiber paper, and ΔV is the voltage change (V), which excluded the internal resistance drop in the discharge period.

2.6. Characterization

Thermogravimetric (TG) analyses were carried out under a nitrogen flux of 60 mL min^{-1} at a heating rate of 10 °C min^{-1} (TGA Q50N2, TA Instruments, New Castle, DE, USA), with approximately 15 mg sample placed in the platinum pan. The surface and cross-section observations were performed using field-emission scanning electron microscopy (FE-SEM) (SU-8020, Hitachi High-Tech Science Corp., Tokyo, Japan). The acceleration voltage was set at 2 kV. Before the FE-SEM observation, platinum sputtering was performed on the samples at 20 mA for 10 s. The nitrogen physisorption measurements were operated at 77 K (NOVA 4200e, Quantachrome Instruments, Kanagawa, Japan). Brunauer–Emmett–Teller (BET) analyses were carried out at relative pressures ranging from 0.01 to 0.3. X-ray diffraction (XRD) analysis was performed by means of an Ultima IV X-ray diffractometer (Rigaku Corp., Tokyo, Japan) with Ni-filtered Cu Kα radiation (1.5418 Å) at 40 kV and 40 mA, and the scanning angle (2θ) was from 5° to 80° at the scanning rate of 1° min^{-1}. Raman spectra were recorded at a laser wavelength and a power of 532 nm and 0.1 mW, respectively (RAMAN-touch VIS-NIR-OUN, Nanophoton Corp., Osaka, Japan). Elemental analysis was operated by a JM10 instrument (J-Science Lab Co., Ltd., Kyoto, Japan). The volume resistiv-ity measurements were performed by a four-probe resistivity meter (Loresta-GP, MCP-T610, Mitsubishi Chemical Analytech Co., Ltd., Tokyo, Japan); three pieces of the pyrolyzed polydopamine-doped cellulose nanofiber papers were prepared for each polydopamine content, and three different positions were measured for each paper sample.

3. Results

3.1. Pyrolysis of Polydopamine-Doped Cellulose Nanofiber Paper

Polydopamine doping and the pyrolysis of the cellulose nanofiber paper were performed as shown in Figure 1. In brief, different amounts of dopamine were first added to the cellulose nanofiber suspension for in situ polymerization to polydopamine. Then, the black-colored aqueous dispersion of the polydopamine-doped cellulose nanofibers was dewatered by suction filtration and solvent exchange, and the samples were then freeze dried. As such, cellulose nanofiber papers with different polydopamine contents (3.4–8.2 wt%) were obtained (Supplementary Figure S1) and pyrolyzed at 700 °C.

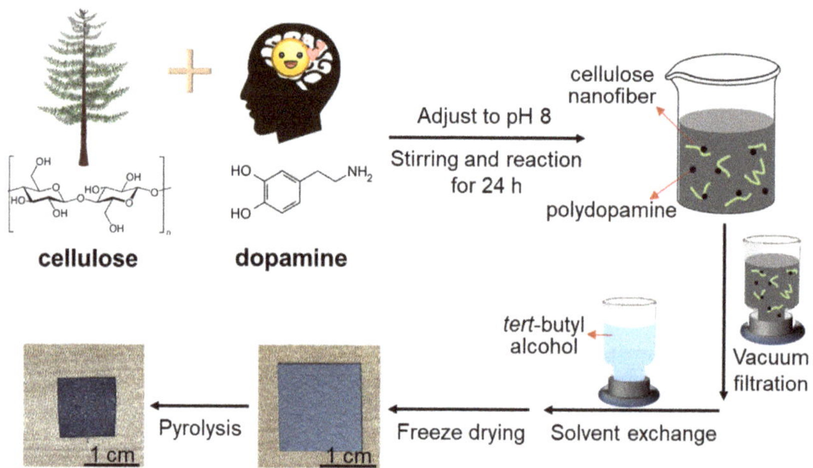

Figure 1. Schematic illustration of the sequential procedure for polydopamine doping and pyrolysis of the cellulose nanofiber paper.

As shown in Figure 2a, the original and polydopamine-doped cellulose nanofiber papers shrank and became brittle to some degree after pyrolysis, but the papers were freestanding, which enabled them to be handled easily for characterization and application testing. Notably, polydopamine doping improved the volume retention (from 15.0% to 23.6%) (Figure 2b), weight retention (from 8.6% to 14.6%) (Figure 2c), and carbon yield (from 16.8% to 28.9%) (Figure 2d, see also Figure S2) of the pyrolyzed cellulose nanofiber paper. Moreover, the weight retention and carbon yield of the pyrolyzed polydopamine-doped cellulose nanofiber paper were higher than the estimated ones. For example, the weight retention and carbon yield of the 4.3 wt% polydopamine-doped cellulose nanofiber paper after pyrolysis were 13.0% and 26.4%, respectively, which were higher than those (10.9% and 19.8%, respectively) estimated from the original cellulose nanofiber paper and neat polydopamine after pyrolysis. Because the polydopamine content was low, it can be expected that the increased weight retention and carbon yield are mainly due to the weight retention of cellulose.

To explain the improved weight and volume retention by polydopamine doping, the TG and derivative thermogravimetric (DTG) curves were analyzed. As it can be seen in Figure 2e, the thermal decomposition temperature (5% weight decrease in the TG curves [35]) of the original cellulose nanofiber paper increased after polydopamine doping from 268.3 °C (polydopamine: 0 wt%) to 282.8 (3.4 wt%), 279.3 (4.3 wt%), and 277.8 °C (8.2 wt%), while that of neat polydopamine was 220.2 °C. The DTG_{peak} temperature and derivative weight of the cellulose nanofiber paper decreased when the polydopamine content increased (Figure 2f). These results suggested that polydopamine doping improves the thermal stability of the cellulose nanofiber paper, thereby affording the increased weight yield and volume. The improved thermal stability can be ascribed to the radical scavenging effect of the catecholic compounds (e.g., polydopamine [36] and melanin [37]). While free radicals, which are generated from cellulose during pyrolysis, unfavorably generate gaseous compounds to cause large weight loss [38], polydopamine can scavenge some of these free radicals and can suppress the weight loss.

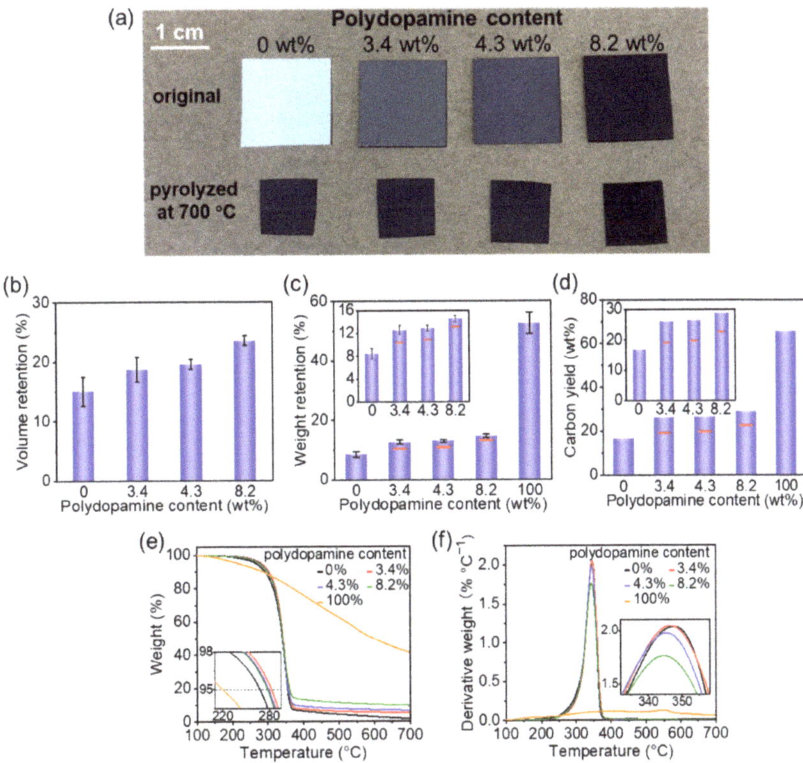

Figure 2. Pyrolysis of the polydopamine-doped cellulose nanofiber paper. (**a**) Cellulose nanofiber papers with different polydopamine contents before and after pyrolysis at 700 °C; (**b**) volume retention and (**c**) weight retention (the red lines indicate the weight retention estimated from the original cellulose and neat polydopamine) and (**d**) carbon yield (the red lines indicate the carbon yield estimated from the pyrolyzed cellulose and pyrolyzed polydopamine) of the pyrolyzed cellulose nanofiber papers with different polydopamine contents; (**e**) thermogravimetric and (**f**) derivative thermogravimetric curves of the cellulose nanofiber papers with different polydopamine contents and neat polydopamine.

3.2. 3D Porous Nanostructures of Pyrolyzed Polydopamine-Doped Cellulose Nanofiber Paper

The 3D porous nanostructures provide unique properties, including high surface areas and permeability, which are beneficial for a variety of applications including energy storage. Therefore, the 3D porous nanostructures of the original and the pyrolyzed polydopamine-doped cellulose nanofiber papers were analyzed accordingly (Figure 3). The undoped cellulose nanofiber paper with a thickness of approximately 270 μm contained a porous nanofiber network (Figure 3a) and a layered structure (Figure S3), comprising the 3D porous nanostructures. The occurrence of these 3D porous nanostructures can be ascribed to the solvent exchange from water to *tert*-butyl alcohol during the paper fabrication process because the *tert*-butyl alcohol with low surface tension suppressed the aggregation of the cellulose nanofibers upon drying [27].

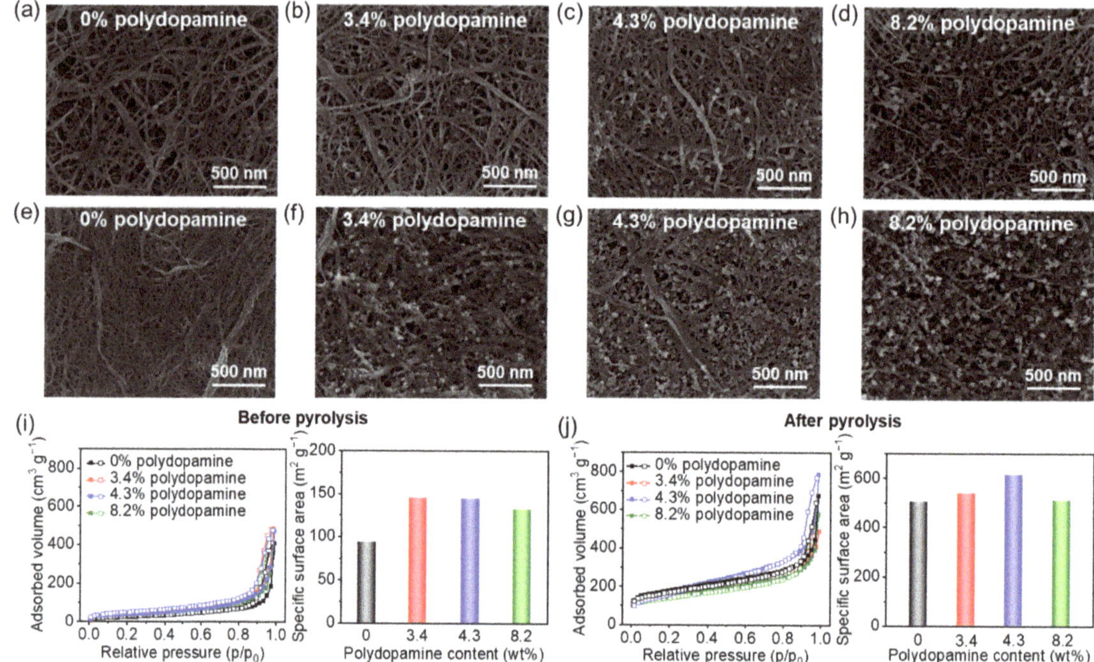

Figure 3. Three-dimensional porous nanostructures of pyrolyzed polydopamine-doped cellulose nanofiber papers. Field emission scanning electron microscopy images of the cellulose nanofiber papers with different polydopamine content (**a–d**) before and (**e–h**) after pyrolysis at 700 °C; N_2 adsorption and desorption isotherms of cellulose nanofiber papers with different polydopamine content and the corresponding specific surface areas (**i**) before and (**j**) after pyrolysis at 700 °C.

Further, the 3D porous nanostructures were maintained even after polydopamine doping (Figure 3b–d and Figure S3). The polydopamine particles formed by in situ polymerization in the cellulose nanofibers had a diameter $\leq$~50 nm (Figure 3b–d), which was much smaller than the neat polydopamine nanoparticles prepared in the absence of cellulose nanofibers (Figure S4), suggesting that cellulose nanofibers can restrict the growth and aggregation of the polydopamine nanoparticles. Owing to the nanosized polydopamine particles, the polydopamine-doped cellulose nanofiber papers had a higher specific surface area (132–145 m^2 g^{-1}) than the original (94 m^2 g^{-1}) (Figure 3i).

It was also confirmed that the polydopamine-doped cellulose nanofiber papers maintained their 3D porous nanostructures after pyrolysis at 700 °C (Figure 3e–h and Figure S3). The thicknesses of the pyrolyzed polydopamine-doped cellulose nanofiber papers were approximately 108, 112, 129, and 142 μm at a polydopamine content of 0%, 3.4%, 4.3%, 8.2%, respectively. The N_2 adsorption and desorption isotherms of the pyrolyzed polydopamine-doped cellulose nanofiber papers indicated an obvious increase of N_2 adsorption in the low relative pressure range and a hysteresis loop in the high relative pressure range, suggesting the presence of micropores (<2 nm) and mesopores (2–50 nm, characteristic of type IV isotherm), respectively [39,40]. The 4.3 wt% polydopamine-doped cellulose nanofiber paper showed the highest specific surface area (617 m^2 g^{-1}) after pyrolysis, which was much larger than that without polydopamine doping (506 m^2 g^{-1}). However, at 8.2 wt% polydopamine, a lower specific surface area (510 m^2 g^{-1}) after pyrolysis was observed, suggesting that the excess amount of polydopamine decreases the resulting specific surface area. The appropriate amount of polydopamine nanoparticles within the cellulose nanofiber paper partially restrain the shrinkage of the porous nanofiber networks upon pyrolysis and provide higher volume retention and specific surface areas (see also

Figure 2b), while an excess amount of polydopamine nanoparticles may block the porous nanostructures, thereby decreasing the specific surface areas. Our results indicate that 3D porous nanostructures with high specific surface areas can be obtained after the pyrolysis of appropriately polydopamine-doped cellulose nanofiber paper.

3.3. Molecular Structure and Electrical Conductivity of Pyrolyzed Polydopamine-Doped Cellulose Nanofiber Papers

The chemical structures and electrical properties of the pyrolyzed polydopamine-doped cellulose nanofiber papers were also analyzed (Figure 4). The Raman spectra showed the G band at approximately 1580 cm^{-1} and D band at approximately 1340 cm^{-1}, which are ascribed to the graphitic carbon domains and the disordered graphitic carbon structures (e.g., edge of graphitic domains and in-plane imperfections) [41], respectively (Figure 4a). This indicates the formation of graphitic carbon structures with disordered regions, such as oxygen- and nitrogen-doped carbon structures, as a result of pyrolysis (see also Figure S2). The graphitic structures after pyrolysis were also confirmed by the XRD spectra. Broad peaks at approximately 23° and 43° (Figure 4b) were observed, which are assigned to the (002) and (10) bands of graphite, respectively [42,43]. The crystallite sizes in the in-plane (L_a) and stacking (L_c) directions of the graphitic carbon can be estimated using Scherrer's formula from the (10) and (002) lattice planes, respectively [41,43]. As shown in Figure 4c, the graphitic carbon domains increased their average width (L_a) from ~1.9 to 2.4 nm (interplanar distance da = ~0.2 nm) with increasing polydopamine content, while their average thickness (L_c) was almost constant at ~1.0 nm. Because neat polydopamine pyrolyzed at the same temperature (700 °C) showed a lower average width of ~1.7 nm than the pyrolyzed polydopamine-doped cellulose nanofiber papers, it was suggested that the combination of polydopamine and cellulose nanofibers can promote the growth of the graphitic carbon domains in the in-plane direction after pyrolysis. Such promoted growth of electrically conductive graphitic carbon domains also contributed toward the increased electrical conductivity (decreased volume resistivity); the volume resistivities of the pyrolyzed polydopamine-doped cellulose nanofiber papers decreased from 8.8 to 5.5 Ω cm when the polydopamine content increased (Figure 4d), indicating high electrical conductivity and the positive effect of polydopamine doping.

3.4. Application as an Electrode for a Supercapacitor

Tuning the porous structure and molecular structure of cellulose-derived nanocarbons by doping is effective in enhancing their energy storage performance [44–46]. To demonstrate the significance of the polydopamine doping and pyrolysis of the cellulose nanofiber paper, the pyrolyzed polydopamine-doped cellulose nanofiber paper was applied as a supercapacitor electrode in a 6 M KOH aqueous electrolyte with a three-electrode system. The previously reported cellulose-derived porous carbons frequently required additives, such as conductive carbon and binder [47–49]. On the other hand, our pyrolyzed polydopamine-doped cellulose nanofiber papers were successfully applied as conductive and free-standing electrodes. Such free-standing property can ensure sufficient transportation of electrons and electrolyte ions between the current collector and the electrodes [50]. The pyrolyzed polydopamine-doped cellulose nanofiber papers showed a larger surrounding area of the CV curve and longer discharge time in the galvanostatic charge/discharge curve than those without polydopamine doping, indicating the higher capacitance of the pyrolyzed polydopamine-doped cellulose nanofiber papers (Figure 5a,b). The specific capacitance values were plotted as a function of the current densities based on the charge-discharge curves (Figure 5c). Notably, the pyrolyzed 4.3 wt% polydopamine-doped cellulose nanofiber paper showed the highest specific capacitance of up to approximately 200 F g^{-1} at current densities of 0.5−20 A g^{-1}, which was considerably higher than that without polydopamine doping (approximately 160 F g^{-1}).

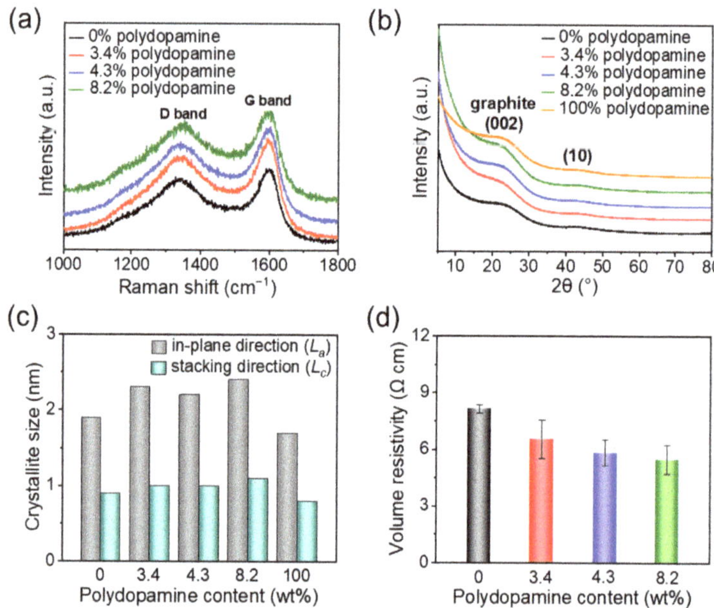

Figure 4. Molecular structure and electrical conductivity of the polydopamine-doped cellulose nanofiber papers pyrolyzed at 700 °C. (**a**) Raman spectra, (**b**) X-ray diffraction spectra, (**c**) crystallite sizes of the graphitic carbon domains in the in-plane (L_a) and stacking (L_c) directions, and (**d**) volume resistivity.

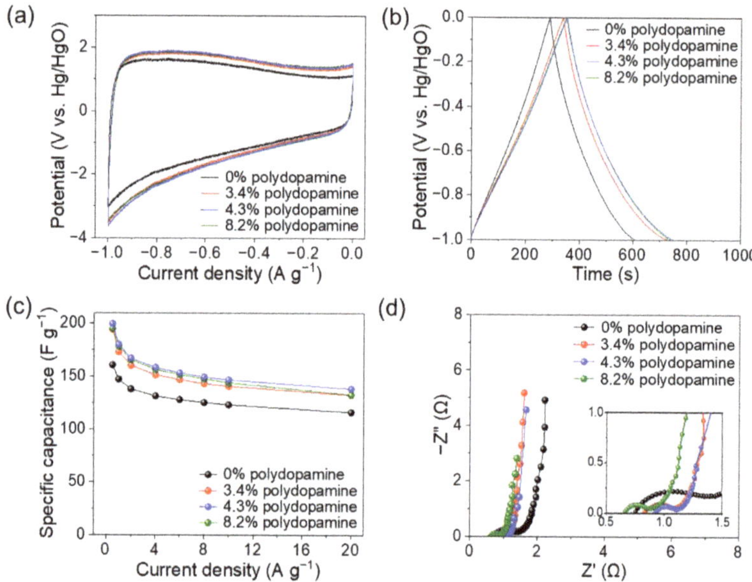

Figure 5. Electrochemical tests of the polydopamine-doped cellulose nanofiber papers pyrolyzed at 700 °C. (**a**) Cyclic voltammetry curves (scan rate: 10 mV s^{-1}); (**b**) galvanostatic charge-discharge curves (current intensity: 0.5 A g^{-1}); (**c**) specific capacitance at the current densities of 0.5−20 A g^{-1}; (**d**) Nyquist plots.

The mechanism that was responsible for the enhanced specific capacitance was discussed. Regardless of polydopamine doping, the pyrolyzed cellulose nanofiber papers presented a rectangular shape with distortion in the CV curves and slight nonlinear charge-discharge curves, which indicate an electric double-layer capacitor behavior with redox pseudocapacitance [47] (Figure 5a,b). The redox pseudocapacitance is derived from disordered carbon structures, such as oxygen- and nitrogen-containing groups [51], which can be estimated as the coulombic efficiency obtained from the ratio of the discharge duration to the charge duration in the charge-discharge curves [52]. The pyrolyzed cellulose nanofiber papers with polydopamine contents of 0, 3.4, 4.3, and 8.2 wt% showed coulombic efficiencies of 111.11, 114.64, 112.96, and 110.76% at 0.5 A g^{-1}, respectively, indicating minimal difference in their redox pseudocapacitance. The resistance of electrolyte ion diffusion within the electrodes and the interfacial charge-transfer at the electrode-electrolyte interface was further analyzed by Nyquist plots, which consist of a linear line at low-frequency region and a semicircle at high-frequency region [47] (Figure 5d). The pyrolyzed cellulose nanofiber papers with and without polydopamine showed a nearly vertical line, representing efficient ionic diffusion [40], which is possibly due to their 3D porous nanostructures (Figure 3e–f). It has been reported that K$^+$ (electrolyte ion) with a large radius can cause the structural deformation of the electrode during the insertion and extraction processes [53,54]. In this study, however, the pyrolyzed polydopamine-doped cellulose nanofiber paper maintained its porous nanostructures after electrochemical tests (Figure S5). Notably, the pyrolyzed polydopamine-doped cellulose nanofiber papers showed a smaller diameter of a semicircle than that without polydopamine doping. This indicates a more efficient interfacial charge-transfer at the electrode-electrolyte interface [40] for the polydopamine-doped samples, owing to their higher electrical conductivity as electrodes (Figure 4d). Thus, the highest specific capacitance of the pyrolyzed 4.3 wt% polydopamine-doped cellulose nanofiber paper is a result of its enhanced electrical conductivity and high specific surface area (Figure 3j). Moreover, the pyrolyzed polydopamine-doped cellulose nanofiber paper provided higher specific capacitance than previously reported cellulose- and other biomass-derived porous nanocarbon materials such as carbonized cellulose aerogel [46], wood-derived carbon nanofiber aerogel [49], and hybrid nanocellulose derived hierarchical porous carbon film [55], suggesting its great potential as promising electrodes in supercapacitors (Table S1).

4. Conclusions

We have demonstrated that the polydopamine doping and pyrolysis of cellulose nanofiber paper can be a promising method that can be used to fabricate the 3D porous nanocarbons with improved carbon yield and volume retention. The pyrolyzed polydopamine-doped cellulose nanofiber paper also offered higher specific surface areas and electrical conductivity than that without polydopamine, thereby affording higher specific capacitance of up to 200 F g^{-1}. The specific capacitance of the polydopamine-doped cellulose nanofiber paper was superior to the cellulose-derived nanocarbons that have been reported previously. Our results suggest that this method will facilitate the effective fabrication of high-performance cellulose-derived 3D porous nanocarbons for energy storage and other electronic applications. This method can be extended to other bionanomaterials, paving the way for future sustainable electronics.

Supplementary Materials: The following are available online at https://www.mdpi.com/article/10.3390/nano11123249/s1, Figure S1: Elemental content of cellulose nanofiber papers with varying dopamine content, Figure S2: Elemental content of the original and the polydopamine-doped cellulose nanofiber papers pyrolyzed at 700 °C, Figure S3: Cross-sectional field emission scanning electron microscopy images of the original and the pyrolyzed cellulose nanofiber paper with different polydopamine content (0% and 4.3%), Figure S4: Field emission scanning electron microscopy images of neat polydopamine, Figure S5: Field emission scanning electron microscopy images of electrode (pyrolyzed 4.3% polydopamine-doped cellulose nanofiber paper) before and after electrochemical tests, Table S1: Specific capacitance values of cellulose-derived porous carbon materials.

Author Contributions: Data curation, L.Z.; formal analysis, L.Z., K.U. and H.K.; funding acquisition, L.Z. and H.K.; investigation, L.Z.; methodology, L.Z. and H.K.; supervision, M.N.; writing—original draft, L.Z.; writing—Review and editing, K.U., M.N. and H.K. All authors have read and agreed to the published version of the manuscript.

Funding: This work was partially supported by Grant-in-Aid for JSPS fellow (JP20J11624 to L.Z.), and by "Nanotechnology Platform Project (Nanotechnology Open Facilities in Osaka University)" from the Ministry of Education, Culture, Sports, Science and Technology, Japan (No. JPMXP09S21OS0029 to H.K.), the Japan Prize Heisei Memorial Research Grant Program (H.K.), and the JST FOREST Program (JPMJFR2003 to H.K.).

Institutional Review Board Statement: Not applicable.

Informed Consent Statement: Not applicable.

Data Availability Statement: The data is included in the main text and/or the Supplementary Materials.

Acknowledgments: Luting Zhu thanks Tsuyoshi Matsuzaki, Comprehensive Analysis Center, SANKEN (The Institute of Scientific and Industrial Research), Osaka University, for the elemental analysis.

Conflicts of Interest: The authors declare no conflict of interest.

References

1. Zhang, L.; Shi, Y.; Wang, Y.; Shiju, N.R. Nanocarbon catalysts: Recent understanding regarding the active sites. *Adv. Sci.* **2020**, *7*, 1902126. [CrossRef]
2. Xin, S.; Guo, Y.; Wan, L. Nanocarbon networks for advanced rechargeable lithium batteries. *Acc. Chem. Res.* **2012**, *45*, 1759–1769. [CrossRef]
3. Mao, X.; Rutledge, G.C.; Hatton, T.A. Nanocarbon-based electrochemical systems for sensing, electrocatalysis, and energy storage. *Nano Today* **2014**, *9*, 405–432. [CrossRef]
4. Dutta, S.; Bhaumik, A.; Wu, K.C.-W. Hierarchically porous carbon derived from polymers and biomass: Effect of interconnected pores on energy applications. *Energy Environ. Sci.* **2014**, *7*, 3574–3592. [CrossRef]
5. Chen, L.; Feng, Y.; Liang, H.; Wu, Z.; Yu, S. Macroscopic-scale three-dimensional carbon nanofiber architectures for electrochemical energy storage devices. *Adv. Energy Mater.* **2017**, *7*, 1700826. [CrossRef]
6. Lv, T.; Liu, M.; Zhu, D.; Gan, L.; Chen, T. Nanocarbon-based materials for flexible all-solid-state supercapacitors. *Adv. Mater.* **2018**, *30*, 1705489. [CrossRef]
7. Liu, L.; Niu, Z.; Chen, J. Unconventional supercapacitors from nanocarbon-based electrode materials to device configurations. *Chem. Soc. Rev.* **2016**, *45*, 4340–4363. [CrossRef]
8. Tu, Y.; Deng, D.; Bao, X. Nanocarbons and their hybrids as catalysts for non-aqueous lithium-oxygen batteries. *J. Energy Chem.* **2016**, *25*, 957–966. [CrossRef]
9. Ma, L.; Li, J.; Li, Z.; Ji, Y.; Mai, W.; Wang, H. Ultra-stable potassium ion storage of nitrogen-doped carbon nanofiber derived from bacterial cellulose. *Nanomaterials* **2021**, *11*, 1130. [CrossRef] [PubMed]
10. Li, J.; Qin, W.; Xie, J.; Lei, H.; Zhu, Y.; Huang, W.; Xu, X.; Zhao, Z.; Mai, W. Sulphur-doped reduced graphene oxide sponges as high-performance free-standing anodes for K-ion storage. *Nano Energy* **2018**, *53*, 415–424. [CrossRef]
11. Titirici, M.-M.; White, R.J.; Brun, N.; Budarin, V.L.; Su, D.S.; del Monte, F.; Clark, J.H.; MacLachlan, M.J. Sustainable carbon materials. *Chem. Soc. Rev.* **2015**, *44*, 250–290. [CrossRef]
12. Inagaki, M.; Yang, Y.; Kang, F. Carbon nanofibers prepared via electrospinning. *Adv. Mater.* **2012**, *24*, 2547–2566. [CrossRef]
13. Spörl, J.M.; Beyer, R.; Abels, F.; Cwik, T.; Müller, A.; Hermanutz, F.; Buchmeiser, M.R. Cellulose-derived carbon fibers with improved carbon yield and mechanical properties. *Macromol. Mater. Eng.* **2017**, *302*, 1700195. [CrossRef]
14. Fox, B. Making stronger carbon-fiber precursors. *Science* **2019**, *366*, 1314–1315. [CrossRef]
15. Trache, D.; Thakur, V.K.; Boukherroub, R. Cellulose nanocrystals/graphene hybrids—a promising new class of materials for advanced applications. *Nanomaterials* **2020**, *10*, 1523. [CrossRef] [PubMed]
16. Rana, A.K.; Frollini, E.; Thakur, V.K. Cellulose nanocrystals: Pretreatments, preparation strategies, and surface functionalization. *Int. J. Biol. Macromol.* **2021**, *182*, 1554–1581. [CrossRef]
17. Xu, P.; Tong, J.; Zhang, L.; Yang, Y.; Chen, X.; Wang, J.; Zhang, S. Dung beetle forewing-derived nitrogen and oxygen self-doped porous carbon for high performance solid-state supercapacitors. *J. Alloys Compd.* **2022**, *892*, 162129. [CrossRef]
18. Wang, Z.; Zhang, X.; Liu, X.; Zhang, Y.; Zhao, W.; Li, Y.; Qin, C.; Bakenov, Z. High specific surface area bimodal porous carbon derived from biomass reed flowers for high performance lithium-sulfur batteries. *J. Colloid Interface Sci.* **2020**, *569*, 22–33. [CrossRef]
19. Zhang, Y.; Wu, C.; Dai, S.; Liu, L.; Zhang, H.; Shen, W.; Sun, W.; Li, C.M. Rationally tuning ratio of micro- to meso-pores of biomass-derived ultrathin carbon sheets toward supercapacitors with high energy and high power density. *J. Colloid Interface Sci.* **2022**, *606*, 817–825. [CrossRef] [PubMed]

20. Zhan, Y.; Bai, J.; Guo, F.; Zhou, H.; Shu, R.; Yu, Y.; Qian, L. Facile synthesis of biomass-derived porous carbons incorporated with CuO nanoparticles as promising electrode materials for high-performance supercapacitor applications. *J. Alloys Compd.* **2021**, *885*, 161014. [CrossRef]

21. Li, T.; Chen, C.; Brozena, A.H.; Zhu, J.Y.; Xu, L.; Driemeier, C.; Dai, J.; Rojas, O.J.; Isogai, A.; Wågberg, L.; et al. Developing fibrillated cellulose as a sustainable technological material. *Nature* **2021**, *590*, 47–56. [CrossRef] [PubMed]

22. Yano, H.; Sasaki, S.; Shams, I.; Abe, K.; Date, T. Wood pulp-based optically transparent film: A paradigm from nanofibers to nanostructured fibers. *Adv. Opt. Mater.* **2014**, *2*, 231–234. [CrossRef]

23. Abe, K.; Iwamoto, S.; Yano, H. Obtaining cellulose nanofibers with a uniform width of 15 nm from wood. *Biomacromolecules* **2007**, *8*, 3276–3278. [CrossRef]

24. Saito, T.; Kimura, S.; Nishiyama, Y.; Isogai, A. Cellulose nanofibers prepared by TEMPO-mediated oxidation of native cellulose. *Biomacromolecules* **2007**, *8*, 2485–2491. [CrossRef] [PubMed]

25. Kondo, T.; Kose, R.; Naito, H.; Kasai, W. Aqueous counter collision using paired water jets as a novel means of preparing bio-nanofibers. *Carbohydr. Polym.* **2014**, *112*, 284–290. [CrossRef]

26. Toivonen, M.S.; Kaskela, A.; Rojas, O.J.; Kauppinen, E.I.; Ikkala, O. Ambient-dried cellulose nanofibril aerogel membranes with high tensile strength and their use for aerosol collection and templates for transparent, flexible devices. *Adv. Funct. Mater.* **2015**, *25*, 6618–6626. [CrossRef]

27. Koga, H.; Namba, N.; Takahashi, T.; Nogi, M.; Nishina, Y. Renewable wood pulp paper reactor with hierarchical micro/nanopores for continuous-flow nanocatalysis. *ChemSusChem* **2017**, *10*, 2560–2565. [CrossRef]

28. Frank, E.; Steudle, L.M.; Ingildeev, D.; Spörl, J.M.; Buchmeiser, M.R. Carbon fibers: Precursor systems, processing, structure, and properties. *Angew. Chem. Int. Ed.* **2014**, *53*, 5262–5298. [CrossRef]

29. Cho, D.; Kim, J.M.; Kim, D. Phenolic resin infiltration and carbonization of cellulose-based bamboo fibers. *Mater. Lett.* **2013**, *104*, 24–27. [CrossRef]

30. Barron, A.B.; Søvik, E.; Cornish, J.L. The roles of dopamine and related compounds in reward-seeking behavior across animal phyla. *Front. Behav. Neurosci.* **2010**, *4*, 163. [CrossRef]

31. Wang, Y.; Zhu, L.; You, J.; Chen, F.; Zong, L.; Yan, X.; Li, C. Catecholic coating and silver hybridization of chitin nanocrystals for ultrafiltration membrane with continuous flow catalysis and gold recovery. *ACS Sustain. Chem. Eng.* **2017**, *5*, 10673–10681. [CrossRef]

32. Jiang, Q.; Derami, H.G.; Ghim, D.; Cao, S.; Jun, Y.S.; Singamaneni, S. Polydopamine-filled bacterial nanocellulose as a biodegradable interfacial photothermal evaporator for highly efficient solar steam generation. *J. Mater. Chem. A* **2017**, *5*, 18397–18402. [CrossRef]

33. Zeng, L.; Zhao, S.; He, M. Macroscale porous carbonized polydopamine-modified cotton textile for application as electrode in microbial fuel cells. *J. Power Sources* **2018**, *376*, 33–40. [CrossRef]

34. Zhu, L.; Huang, Y.; Morishita, Y.; Uetani, K.; Nogi, M.; Koga, H. Pyrolyzed chitin nanofiber paper as a three-dimensional porous and defective nanocarbon for photosensing and energy storage. *J. Mater. Chem. C* **2021**, *9*, 4444–4452. [CrossRef]

35. Fukuzumi, H.; Saito, T.; Okita, Y.; Isogai, A. Thermal stabilization of TEMPO-oxidized cellulose. *Polym. Degrad. Stab.* **2010**, *95*, 1502–1508. [CrossRef]

36. Cho, J.H.; Vasagar, V.; Shanmuganathan, K.; Jones, A.R.; Nazarenko, S.; Ellison, C.J. Bioinspired catecholic flame retardant nanocoating for flexible polyurethane foams. *Chem. Mater.* **2015**, *27*, 6784–6790. [CrossRef]

37. Shanmuganathan, K.; Cho, J.H.; Iyer, P.; Baranowitz, S.; Ellison, C.J. Thermooxidative stabilization of polymers using natural and synthetic melanins. *Macromolecules* **2011**, *44*, 9499–9507. [CrossRef]

38. Zhang, C.; Chao, L.; Zhang, Z.; Zhang, L.; Li, Q.; Fan, H.; Zhang, S.; Liu, Q.; Qiao, Y.; Tian, Y.; et al. Pyrolysis of cellulose: Evolution of functionalities and structure of bio-char versus temperature. *Renew. Sustain. Energy Rev.* **2021**, *135*, 110416. [CrossRef]

39. Zhang, X.; Ma, L.; Gan, M.; Fu, G.; Jin, M.; Lei, Y.; Yang, P.; Yan, M. Fabrication of 3D lawn-shaped N-doped porous carbon matrix/polyaniline nanocomposite as the electrode material for supercapacitors. *J. Power Sources* **2017**, *340*, 22–31. [CrossRef]

40. Gao, L.; Xiong, L.; Xu, D.; Cai, J.; Huang, L.; Zhou, J.; Zhang, L. Distinctive construction of chitin-derived hierarchically porous carbon microspheres/polyaniline for high-rate supercapacitors. *ACS Appl. Mater. Interfaces* **2018**, *10*, 28918–28927. [CrossRef]

41. Biscoe, J.; Warren, B.E. An X-ray study of carbon black. *J. Appl. Phys.* **1942**, *13*, 364–371. [CrossRef]

42. Jurkiewicz, K.; Pawlyta, M.; Burian, A. Structure of carbon materials explored by local transmission electron microscopy and global powder diffraction probes. *J. Carbon Res.* **2018**, *4*, 68. [CrossRef]

43. Tanaka, K.; Ueda, M.; Koike, T.; Yamabe, T. X-ray diffraction studies of pristine and heavily-doped polyacenic materials. *Synth. Met.* **1988**, *25*, 265–275. [CrossRef]

44. Liu, Q.; Jing, S.; Wang, S.; Zhuo, H.; Zhong, L.; Peng, X.; Sun, R. Flexible nanocomposites with ultrahigh specific areal capacitance and tunable properties based on a cellulose derived nanofiber-carbon sheet framework coated with polyaniline. *J. Mater. Chem. A* **2016**, *4*, 13352–13362. [CrossRef]

45. Liang, H.W.; Wu, Z.Y.; Chen, L.F.; Li, C.; Yu, S.H. Bacterial cellulose derived nitrogen-doped carbon nanofiber aerogel: An efficient metal-free oxygen reduction electrocatalyst for zinc-air battery. *Nano Energy* **2015**, *11*, 366–376. [CrossRef]

46. Li, X.; Zhang, Y.; Zhang, J.; Wang, C. Isolated Fe atoms dispersed on cellulose-derived nanocarbons as an efficient electrocatalyst for the oxygen reduction reaction. *Nanoscale* **2019**, *11*, 23110–23115. [CrossRef]

47. Cai, J.; Niu, H.; Li, Z.; Du, Y.; Cizek, P.; Xie, Z.; Xiong, H.; Lin, T. High-performance supercapacitor electrode materials from cellulose-derived carbon nanofibers. *ACS Appl. Mater. Interfaces* **2015**, *7*, 14946–14953. [CrossRef] [PubMed]
48. Tian, X.; Zhu, S.; Peng, J.; Zuo, Y.; Wang, G.; Guo, X.; Zhao, N.; Ma, Y.; Ma, L. Synthesis of micro- and meso-porous carbon derived from cellulose as an electrode material for supercapacitors. *Electrochim. Acta* **2017**, *241*, 170–178. [CrossRef]
49. Li, S.C.; Hu, B.C.; Ding, Y.W.; Liang, H.W.; Li, C.; Yu, Z.Y.; Wu, Z.Y.; Chen, W.S.; Yu, S.H. Wood-derived ultrathin carbon nanofiber aerogels. *Angew. Chem. Int. Ed.* **2018**, *57*, 7085–7090. [CrossRef]
50. Xu, X.; Shen, J.; Li, F.; Wang, Z.; Zhang, D.; Zuo, S.; Liu, J. Fe_3O_4@C Nanotubes Grown on Carbon Fabric as a Free-Standing Anode for High-Performance Li-Ion Batteries. *Chem. A Eur. J.* **2020**, *26*, 14708–14714. [CrossRef] [PubMed]
51. Pang, J.; Zhang, W.; Zhang, H.; Zhang, J.; Zhang, H.; Cao, G.; Han, M.; Yang, Y. Sustainable nitrogen-containing hierarchical porous carbon spheres derived from sodium lignosulfonate for high-performance supercapacitors. *Carbon* **2018**, *132*, 280–293. [CrossRef]
52. Xia, X.; Hao, Q.; Lei, W.; Wang, W.; Wang, H.; Wang, X. Reduced-graphene oxide/molybdenum oxide/polyaniline ternary composite for high energy density supercapacitors: Synthesis and properties. *J. Mater. Chem.* **2012**, *22*, 8314–8320. [CrossRef]
53. Li, J.; Zhuang, N.; Xie, J.; Zhu, Y.; Lai, H.; Qin, W.; Javed, M.S.; Xie, W.; Mai, W. Carboxymethyl cellulose binder greatly stabilizes porous hollow carbon submicrospheres in capacitive K-Ion storage. *ACS Appl. Mater. Interfaces* **2019**, *11*, 15581–15590. [CrossRef] [PubMed]
54. Xu, X.; Mai, B.; Liu, Z.; Ji, S.; Hu, R.; Ouyang, L.; Liu, J.; Zhu, M. Self-sacrificial template-directed ZnSe@C as high performance anode for potassium-ion batteries. *Chem. Eng. J.* **2020**, *387*, 124061. [CrossRef]
55. Li, Z.; Ahadi, K.; Jiang, K.; Ahvazi, B.; Li, P.; Anyia, A.O.; Cadien, K.; Thundat, T. Freestanding hierarchical porous carbon film derived from hybrid nanocellulose for high-power supercapacitors. *Nano Res.* **2017**, *10*, 1847–1860. [CrossRef]

Communication

Enhancement of Luminance in Powder Electroluminescent Devices by Substrates of Smooth and Transparent Cellulose Nanofiber Films

Shota Tsuneyasu [1], Rikuya Watanabe [1], Naoki Takeda [1], Kojiro Uetani [2], Shogo Izakura [3], Keitaro Kasuya [3], Kosuke Takahashi [3] and Toshifumi Satoh [1,*]

[1] Department of Media Engineering, Graduate School of Engineering, Tokyo Polytechnic University, 1583 Iiyama, Atsugi, Kanagawa 243-0297, Japan; s.tsuneyasu@mega.t-kougei.ac.jp (S.T.); m1716089@st.t-kougei.ac.jp (R.W.); naoki.takeda@toppan.co.jp (N.T.)

[2] The Institute of Scientific and Industrial Research (SANKEN), Osaka University, Mihogaoka 8-1, Ibaraki, Osaka 567-0047, Japan; uetani@eco.sanken.osaka-u.ac.jp

[3] Graduate School of Engineering, Osaka University, Mihogaoka 8-1, Ibaraki, Osaka 567-0047, Japan; u895635a@gmail.com (S.I.); keitaro_k@chem.eng.osaka-u.ac.jp (K.K.); h.takahashi@eco.sanken.osaka-u.ac.jp (K.T.)

* Correspondence: toshi@mega.t-kougei.ac.jp

Abstract: Powder electroluminescent (EL) devices with an electric field type excitation are surface light sources that are expected to have a wide range of practical applications, owing to their high environmental resistance; however, their low luminance has hindered their use. A clarification of the relationship between the properties of the film substrates and the electroluminescence is important to drastically improve light extraction efficiency. In this study, powder EL devices with different substrates of various levels of surface roughness and different optical transmittances were fabricated to quantitatively evaluate the relationships between the substrate properties and the device characteristics. A decrease in the surface roughness of the substrate caused a clear increase in both the current density and the luminance. The luminance was found to have a direct relationship with the optical transmittance of the substrates. The powder EL device, which was based on a cellulose nanofiber film and was the smoothest and most transparent substrate investigated, showed the highest luminance (641 cd/cm^2) when 300 V was applied at 1 kHz.

Keywords: powder electroluminescent device; cellulose nanofiber; paper electronic device; paper-based light-emitting device

Citation: Tsuneyasu, S.; Watanabe, R.; Takeda, N.; Uetani, K.; Izakura, S.; Kasuya, K.; Takahashi, K.; Satoh, T. Enhancement of Luminance in Powder Electroluminescent Devices by Substrates of Smooth and Transparent Cellulose Nanofiber Films. *Nanomaterials* **2021**, *11*, 697. https://doi.org/10.3390/nano11030697

Academic Editors: Takuya Kitaoka and Elena D. Obraztsova

Received: 17 February 2021
Accepted: 8 March 2021
Published: 10 March 2021

Publisher's Note: MDPI stays neutral with regard to jurisdictional claims in published maps and institutional affiliations.

1. Introduction

The development of a flexible flat light-emitting device that includes organic and inorganic light-emitting devices beyond the traditional solid and rigid displays of liquid crystal or light-emitting diodes is useful for the establishment of a next-generation super-smart society. A powder electroluminescent (EL) device is a traditional flexible flat light source that involves inorganic EL particles and can be fabricated using a simple printing system on a film substrate, such as copy paper and plastic films. Compared with the conventional organic flat light-emitting devices with a current-type excitation (which are very sensitive to the surrounding environment (i.e., water and oxygen)), the powder EL device has a high environmental resistance, owing to its characteristic electric field type excitation. Because of this advantage, the powder EL devices, which can be fabricated by 3D printing [1,2], exhibit good mechanical durability, such as bending durability [3], flexibility [4], and stretchability [5,6]. They can also be used as sensors for ammonia [7], carbon dioxide [8], humidity [9], and temperature [10]. Unique functionalities, such as the self-recovery of their structure [11–13] and the generation of sound [14–16], have also been developed through the powder EL system. However, the practical use of powder EL devices has only been based on indirect illumination, owing to their low luminance. For a long time, organic light emitting devices were thought to have difficulty in achieving laser oscillation due to their low brightness. Recently, the laser action demonstrated in organic semiconductors by the distributed feedback structure has exhibited optical confinement effects [17]. If the luminance in powder EL devices with low luminance can be significantly improved, the desired laser action might be observed by introducing the rib and slot-waveguide structures [18,19]. The intrinsic improvement of the luminance for a powder EL device is the key to develop flexible and versatile light-emitting devices.

In terms of achieving high luminance from a powder EL device, researchers have focused their investigations on the phosphor [20] and the high dielectric polymer [21] components. Additionally, a high luminance has been achieved by drastically modifying the device structure [22,23]. However, we previously focused on the surface roughness of the paper substrate and found that the planarization of the substrate is effective to achieve high luminescence from a bottom emission-type powder EL device [24]. Therefore, there is great potential to improve EL luminance by investigating the properties of the film substrates. However, the direct effect of the film properties (which include surface roughness and transparency) on EL luminance has not yet been clarified. This is because EL luminance is complex and is affected by multiple factors, such as applied voltage and frequency. Furthermore, powder EL devices with a paper substrate have often employed a top emission-type structure that is unaffected by the optical transmittance of the substrate films [25–27]. To essentially improve the light extraction efficiency in powder EL devices, the direct effect of the properties of the film substrates needs to be investigated by using a bottom emission-type device.

Recently, cellulose nanofiber (CNF) films have been developed from various phytomass resources and have been shown to exhibit high transparency [28,29] and minimal surface roughness [30,31], as well as good flexibility. Compared with conventional plastics, CNF films are also known to display high strength [32,33] and thermal stability (a low coefficient of thermal expansion) [34,35], owing to the extended chain crystals of natural cellulose type I. This CNF film is a "paper" material in which the CNFs selfagglutinate through the strong hydrogen bonds on their surface. Therefore, the CNF film is proposed to be the most legitimate successor of a conventional paper substrate in powder EL devices, owing to its high performance.

In this study, we aimed to elucidate the effects of surface roughness and the transmittance of the film substrates on the luminance of powder EL devices and achieve a high luminance that exceeds that of the devices composed of conventional paper and plastic film substrates. Powder EL devices were fabricated with four different substrates: tracing paper, polyethylene naphthalate (PEN) film, commercial CNF (C-CNF) sheets, and 2,2,6,6-tetramethylpiperidine-1-oxyl (TEMPO)-oxidized CNF (TO-CNF) films. Each device had a different surface roughness and optical transmittance, and the luminance performances of each device were compared.

2. Materials and Methods

2.1. Materials

Tracing paper (Sakae Technical Paper Co., Ltd., Tokyo, Japan), the C-CNF sheets (WMa-100FM, Sugino Machine Ltd., Toyama, Japan), and the PEN film (Q65FA, Teijin DuPont Films, Tokyo, Japan) were purchased and used as received.

The TO-CNF film was produced according to the previously reported method. In brief, Japanese cedar (*Cryptomeria japonica*) chips (50 g) were dewaxed by immersion in 1 L of acetone overnight. After removing the acetone, the resulting product was bleached in 1.5 L of 1.0 wt.% $NaClO_2$ solution under acidic conditions at 80–90 °C for 6 h until the product became white, in accordance with Wise's method [36]. The product was then washed with distilled water to obtain wood pulp. The pulp (with a dry weight of 5 g) was dispersed in 500 mL of distilled water and 0.08 g of TEMPO with the addition of 0.5 g of NaBr [37]. Oxidation was initiated by adding 27 mL of 1.8 M NaClO aqueous solution, and the pH was kept higher than 10 by adding 0.5 M NaOH. After reacting for 7 h, the oxidation was quenched by adding 5 mL of ethanol and then washed until the pH reached ~7 [38]. The product was redispersed in 500 mL of distilled water, further treated by adding 5.0 g of $NaClO_2$ under acidic conditions, and stirred for 48 h to selectively convert the aldehyde group in the samples to a carboxylate group [39].

After washing, a TEMPO-oxidized pulp with a carboxylate content of 1.67 mmol/g (determined via conductometric titration) was obtained. The TEMPO-oxidized pulp was preliminarily agitated at 24,000 rpm by a high-speed blender for 10 min and was then passed through a high-pressure water jet system (Star Burst 10, HJP-25008, Sugino Machine Ltd., Toyama, Japan) at 180 MPa for 50 cycles to produce an aqueous suspension of TEMPO-oxidized cellulose nanofibers. After removing the bubbles, the viscous nanofiber suspension (with a concentration of 1.5 wt.%) was cast on an acrylic plate by using a metal formwork with internal dimensions of 7×7 cm^2 and a thickness of 2 mm, and this cast was dried in an oven at 55 °C overnight to form the TO-CNF films.

Zinc sulfide-type particles (GG45, Osram Sylvania, Wilmington, MA, USA) were used as a phosphor layer, barium titanium oxide (620-07545, Kishida Chemical Co., Ltd., Osaka, Japan) was used as a dielectric layer, cyanoethyl polyvinyl alcohol (cyanoresin, CR-V, Shinetsu Chemical Co., Ltd., Tokyo, Japan) was used as a high dielectric polymer, poly(2,3-dihydrothieno-1,4-dioxin)–poly(styrenesulfonate) (PEDOT:PSS, 768650, Sigma-Aldrich Co., LLC., Tokyo, Japan) was used as a transparent electrode, and Ag-paste (MP-603S, Mino Group Co., Ltd., Gifu, Japan) was used as a back electrode, all as received. Cyclohexanone (037-05096) was purchased from Fujifilm Wako Pure Chemical Co., Osaka, Japan.

2.2. Characterization of the Substrate Films

The regular light transmittance was measured with a spectrophotometer (U-3900, Hitachi High-Tech Corp., Tokyo, Japan). The haze was evaluated using a haze meter (HZ-V3, Suga Test Instruments Co., Ltd., Tokyo, Japan). The surface roughness (the root mean square (RMS)) was determined for an area of 20×20 μm^2 using an atomic force microscope (AFM, Nanocute, SII Nano Technology Inc., Chiba, Japan) in dynamic force mode [40,41]. The three-dimensionally extended RMS of cross-sectional curves defined in JIS B 0601 (ISO 4287) was calculated for the measured surfaces, and it is expressed as the square root of the average of the square of deviation from the reference surface to the specified surface. In this study, RMS was calculated for the entire 20×20 μm^2 angle of view.

2.3. Fabrication of the EL Devices

The cyanoresin was mixed in cyclohexanone at 30 wt.% to prepare the high-dielectric polymer paste. Then, 40 wt.% of a zinc sulfide-type phosphor was dispersed in the polymer paste. Barium titanium oxide was also dispersed in the polymer paste, in the same weight ratio as the phosphor had been dispersed. The transparent electrode, phosphor layer, dielectric layer, and back electrode were laminated by an automatic screen-printing machine (TU2020-C, Seritech Co., Ltd., Osaka, Japan) on each substrate in this order. Each layer was individually dried in an oven at 80 °C for 6 min; the transparent electrode layer was dried for 10 min at 80 °C to fully evaporate the extra solvents. The light-emitting area of the devices was set as 1.0×1.0 cm^2. The cross-sectional layer structure for each EL device was observed by an optical microscope (WPA-micro, Photonic Lattice Inc., Miyagi, Japan).

2.4. EL Measurements

The voltage dependences of the current and the luminance of the devices were determined by using an EL measurement system (SX-1152, Iwatsu Electric Co., Ltd., Tokyo, Japan).

3. Results and Discussion

The regular light transmittance of the four films was investigated. Figure 1a–e shows a photograph of and the optical transmittance of each film. The TO-CNF, PEN, and C-CNF films appeared highly transparent, as shown in Figure 1a–c, respectively; however, each of these films had a different transmittance spectrum (Figure 1e). The TO-CNF film showed the highest transparency among the four, and the transmittance in the visible light range for this film reached approximately 90%. The PEN film showed almost the same transmittance as the TO-CNF film at wavelengths greater than 700 nm, but the transmittance gradually decreased at lower wavelengths. The C-CNF film had a transmittance of ~30% over a wide wavelength range. The tracing paper was completely opaque with the lowest transmittance of below 3%, as shown in Figure 1d, but it has long been used as a powder EL substrate in previous studies [10,24].

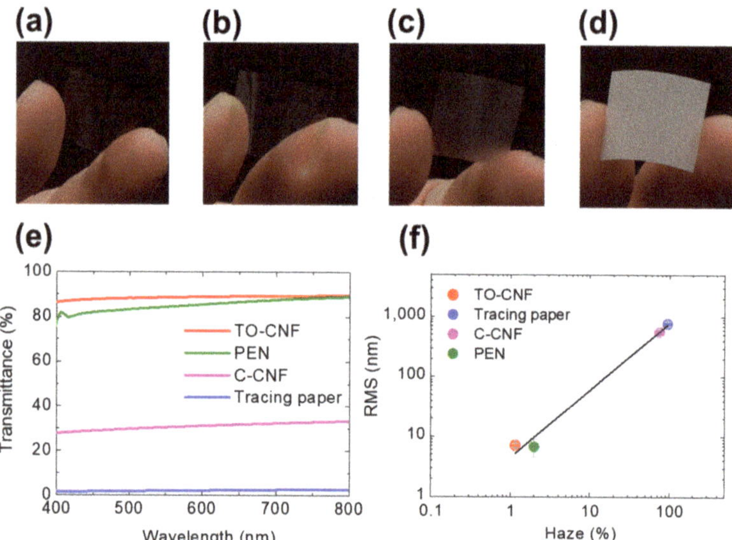

Figure 1. Photographs of (**a**) 2,2,6,6-tetramethylpiperidine-1-oxyl (TEMPO)-oxidized cellulose nanofiber (CNF) (TO-CNF), (**b**) polyethylene naphthalate (PEN), (**c**) commercial CNF (C-CNF), and (**d**) tracing paper films. (**e**) Transmittance spectra of each film. (**f**) The relationship between root mean square (RMS) roughness and haze.

We evaluated the surface roughness of each film using the RMS values obtained from the AFM measurement (Figure 2). The RMS roughness of each of the TO-CNF, PEN, C-CNF, and tracing paper films were found to be around 6.78 ± 2.3 nm, 7.19 ± 0.54 nm, 560.9 ± 72.7 nm, and 769 ± 64.5 nm, respectively (as shown in Table 1). We found that the film with a smaller RMS value showed a higher transmittance. To evaluate the optical properties, the haze of each film was also measured. The haze of the TO-CNF, PEN, C-CNF, and tracing paper substrates were approximately 1.99 ± 0.06%, 1.15 ± 0.10%, 74.34 ± 0.41%, and 93.96 ± 0.04%, respectively. We obtained a linear relationship between the RMS values and haze, as shown in Figure 1f. The surface roughness of each film was shown as directly related to its optical transparency. The negative linear correlation between RMS and transparency in films was also found for microcrystalline diamond [42]. In general, it is also empirically self-evident that a rough surface leads to opacity, as seen in frosted glass. Regardless of the chemical composition, the transparency of the film is strongly correlated with the surface roughness, which is also shown in this study.

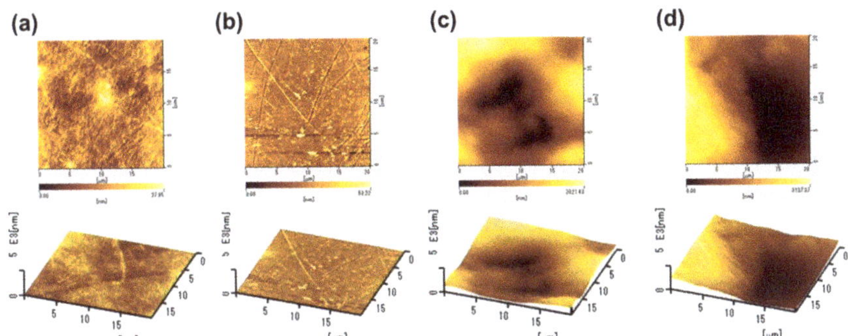

Figure 2. Atomic force microscope (AFM) images of (**a**) TO-CNF, (**b**) PEN, (**c**) C-CNF, and (**d**) tracing paper substrates for the calculation of RMS roughness.

Table 1. The characteristics of each film and the thickness of the functional layer on those devices.

	Film Thickness (μm)	RMS Roughness (nm)	Haze (%)	Thickness of the Functional Layers (μm)
TO-CNF	22.0 ± 2.0	6.78 ± 2.3	1.99 ± 0.06	106.0 ± 2.3
PEN	128.0 ± 2.8	7.19 ± 0.54	1.15 ± 0.10	129.4 ± 3.8
C-CNF	54.0 ± 0.4	560.9 ± 72.7	74.34 ± 0.41	94.0 ± 3.6
Tracing paper	82.0 ± 1.4	769 ± 64.5	93.96 ± 0.04	122.6 ± 6.7

The powder EL devices were fabricated using these films as substrates. To explore the electrode distance of each device, the cross-section of each EL device was observed using an optical microscope (Figure S1). The thickness of the functional layers of the devices with the TO-CNF, PEN, C-CNF, and tracing paper substrates were measured to be 106.0 ± 2.3 μm, 129.4 ± 3.8 μm, 94.0 ± 3.6 μm, and 122.6 ± 6.7 μm, respectively, although the thicknesses of the film substrates were significantly different. It was therefore suggested that almost the same electric field intensity could be obtained when the same voltage was applied, regardless of the type of substrate film.

Figure 3a–d shows photographs of the luminescence of each EL device with a different substrate film under the application of an AC voltage of ±170 V at 1.2 kHz. All of the EL devices emitted blue light. The emissions from the devices with C-CNF and tracing paper substrates were uneven and dark. This was proposed to arise from the variation of the electrode distance, which could not be detected from local observation with an optical microscope. The variation of the electrode distance occurred owing to the large surface roughness of the substrate, which led to a variation in the electric field intensity. The EL spectrum of these devices was investigated to ascertain the effect of the substrates on EL emission. The EL intensity of the devices with the TO-CNF and PEN substrates under application of the same voltage was higher than those with the C-CNF and tracing paper substrates (Figure 3e). This was considered to arise from the differences in the surface roughness and optical transmittance of the substrates. As shown in the inset of Figure 3e, the EL band with an emission peak was observed at a wavelength of 490 nm, which was almost identical to that observed from previous devices [24].

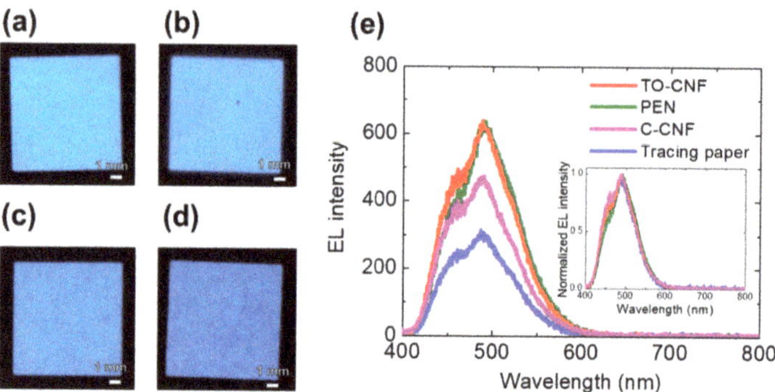

Figure 3. Photographs of the powder EL devices with a substrate of (**a**) TO-CNF, (**b**) PEN, (**c**) C-CNF, and (**d**) tracing paper under the application of an AC voltage of ±170 V at 1.2 kHz. (**e**) Electroluminescent (EL) spectra of the devices with each substrate under the application of an AC voltage of ±170 V at 1.2 kHz. The inset shows the normalized EL spectra of the devices.

To clarify the EL properties, the voltage dependence of the current density of the EL devices with each substrate was measured (Figure 4a). We set the current density at 1 kHz because the commercialized zinc sulfide-type particles (GG45, Osram Sylvania) used in this study are known to give the best luminescence efficiency at 1 kHz [43]. The current density of all devices increased as the applied voltage increased. The current density of the devices with the TO-CNF and PEN substrates was significantly higher than those with the C-CNF and tracing paper substrates. As shown in Figure 4b, under application of an AC voltage of ±300 V at 1 kHz, the current density exponentially decreased as the RMS roughness of the substrate film was increased. Thus, the current density of the powder EL device was found to be dependent on the RMS roughness of its substrate.

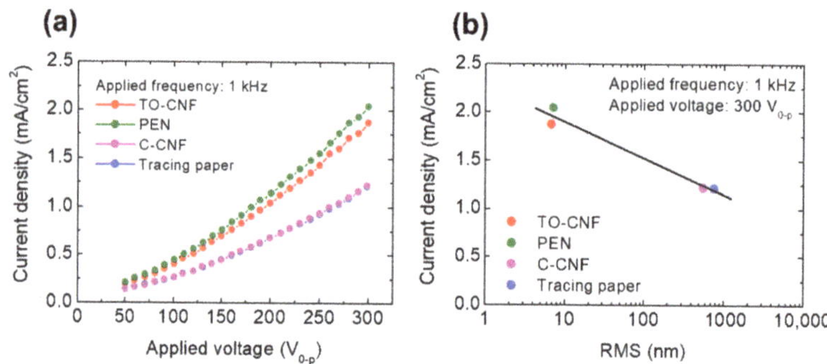

Figure 4. EL property measurement. (**a**) The relationship between current density at an applied frequency of 1 kHz and applied voltage. (**b**) The relationship between current density at an applied voltage of ±300 V_{0-p} at 1 kHz and the RMS roughness of the substrate films.

The applied voltage dependence of the luminance of the EL devices with each substrate was measured (Figure 5a). By scanning the applied voltage from ±50–300 V, the luminance of all devices increased with the increase of the applied voltage. Under the application of ±300 V at 1 kHz, the luminance from the device with the TO-CNF substrate was approximately 641 cd/m^2 and was the highest compared with those of the devices with the other substrates. Compared with previous reports on bottom emission-type EL devices, which showed a maximum luminance of 150 cd/m^2 under application of an AC voltage of $\pm$150 V at 0.4 kHz [44], our 641 cd/m^2 demonstrated a drastic enhancement of the luminance for the bottom emission-type EL device with a paper substrate.

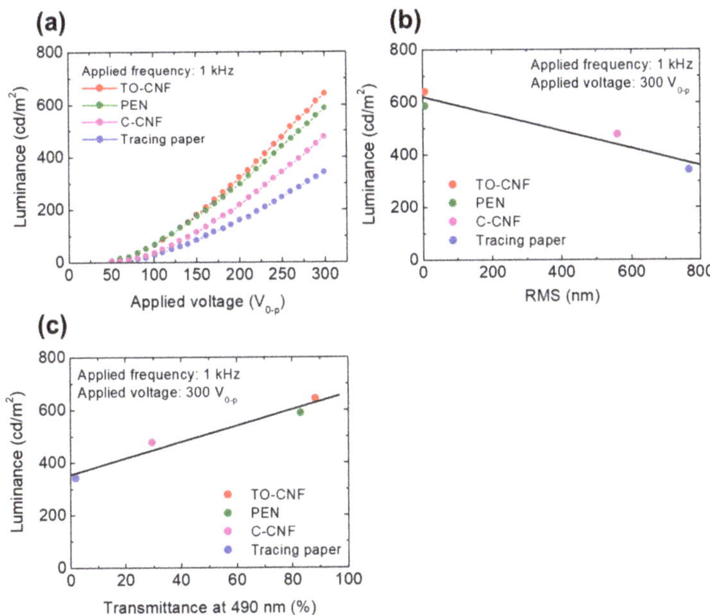

Figure 5. (**a**) The luminance of the EL devices versus the applied voltage characteristics at an applied frequency of 1 kHz. The luminance of the EL devices at an applied voltage of 300 V$_{0-p}$ at 1 kHz versus (**b**) RMS roughness and (**c**) transmittance of each substrate at a wavelength of 490 nm.

To quantitatively elucidate the relationship between the substrate characteristics and the obtained luminance, the RMS roughness of the film substrates and their EL luminance under application of an AC voltage of $\pm$300 V at 1 kHz is plotted in Figure 5b. The luminance linearly increased with a decrease in RMS roughness in a similar manner as the current density described above in Figure 4b. Thus, the current density can be increased by an improvement of the RMS roughness of the substrates, which will lead to a high luminance in powder EL devices. Figure 5c shows the clear relationship between the transmittance of the substrate films at 490 nm (which is the wavelength of the emission peak) and the luminescence of the EL devices. The luminance linearly increased with an increase of the optical transmittance of the substrates. It is now clearly demonstrated that the required traits of a substrate for obtaining high luminance in a powder EL device are not only a smooth surface but also a high transparency. These findings are key factors for suitably designing the luminance in powder EL devices.

4. Conclusions

We have clarified the effect of the physical state of the film substrate (surface roughness and transparency) on the luminance of powder EL devices, which has not been focused on before. At the same time, we have succeeded in producing the powder EL device with the highest luminance of 641 cd/cm^2 when 300 V was applied at 1 kHz by using the substrate of cellulose nanofiber film, a next-generation paper material with high smoothness, transparency, and other excellent properties.

A typical application of powdered EL devices is as a surface emitting light source, and displays and lighting are its most common applications. The present study, which clarified the physical guideline for improving the luminance (the lighting performance) using natural materials that promote the elimination of plastics as the base material, is expected to provide a significant guideline for the development of next-generation high-performance lighting. We believe that the findings of this research will be useful in the future for highly linear point light sources because the device might achieve a laser oscillation by the introduction of rib and slot-waveguides structures.

Supplementary Materials: The following are available online at https://www.mdpi.com/2079-4991/11/3/697/s1, Figure S1: Cross-sectional optical microscopy images of (a) TO-CNF, (b) PEN, (c) C-CNF, and (d) tracing paper substrates.

Author Contributions: Conceptualization, S.T., K.U., and T.S.; methodology, S.T. and K.U.; formal analysis, S.T. and K.U.; investigation, K.T. and K.U.; resources, S.I., K.K., R.W., and N.T.; writing—original draft preparation, S.T. and K.U.; writing—review and editing, T.S.; supervision, T.S.; project administration, S.T. and K.U.; funding acquisition, S.T., K.U., and T.S. All authors have read and agreed to the published version of the manuscript.

Funding: This work was financially supported by "FY2016 MEXT Private University Research Branding Project", Fuji Seal Foundation and JGC-S Scholarship Foundation. This work was performed under the Cooperative Research Program of "Network Joint Research Center for Materials and Devices".

Acknowledgments: We thank the Edanz Group for editing a draft of this manuscript.

Conflicts of Interest: The authors have no conflicts to declare.

References

1. Brubaker, C.D.; Newcome, K.N.; Jennings, G.K.; Adams, D.E. 3D-Printed alternating current electroluminescent devices. *J. Mater. Chem. C* **2019**, *7*, 5573–5578. [CrossRef]
2. Liu, D.; Ren, J.; Wang, J.; Xing, W.; Qian, Q.; Chen, H.; Zhou, N. Customizable and stretchable fibre-shaped electroluminescent devices via mulitcore-shell direct ink writing. *J. Mater. Chem. C* **2020**, *8*, 15092–15098. [CrossRef]
3. Ohsawa, M.; Hashimoto, N.; Takeda, N.; Tsuneyasu, S.; Satoh, T. Flexible powder electroluminescent device on transparent electrode of printed invisible silver-grid laminated with conductive polymer. *Flex. Print. Electron.* **2020**, *5*, 025004. [CrossRef]
4. Ohsawa, M.; Hashimoto, N.; Takeda, N.; Tsuneyasu, S.; Satoh, T. Bending reliability of transparent electrode of printed invisible silver-grid/PEDOT:PSS on flexible epoxy film substrate for powder electroluminescent device. *Microelectron. Reliab.* **2020**, *109*, 113673. [CrossRef]
5. Yang, C.H.; Chen, B.; Zhou, J.; Chen, Y.M.; Suo, Z. Electroluminescence of giant stretchability. *Adv. Mater.* **2016**, *28*, 4480–4484. [CrossRef]
6. Hu, D.; Xu, X.; Miao, J.; Gidron, O.; Meng, H. A Stretchable Alternating Current Electroluminescent Fiber. *Materials* **2018**, *11*, 184. [CrossRef] [PubMed]
7. En-On, J.; Tuantranont, A.; Kerdcharoen, T.; Wongchoosuk, C. Flexible alternating current electroluminescent ammonia gas sensor. *RSC Adv.* **2017**, *7*, 16885–16889. [CrossRef]
8. Seekaew, Y.; Wongchoosuk, C. A novel graphene-based electroluminescent gas sensor for carbon dioxide detection. *Appl. Surf. Sci.* **2019**, *479*, 525–531. [CrossRef]
9. He, Y.; Zhang, M.; Zhang, N.; Zhu, D.; Huang, C.; Kang, L.; Zhang, J. Paper-Based ZnS: Cu Alternating Current Electroluminescent Devices for Current Humidity Sensors with High-Linearity and Flexibility. *Sensors* **2019**, *19*, 4607. [CrossRef] [PubMed]
10. Tsuneyasu, S.; Takeda, N.; Satoh, T. Novel powder electroluminescent device enabling control of emission color by thermal response. *J. Soc. Inf. Disp.* **2021**, *29*, 207–212. [CrossRef]
11. Liang, G.; Liu, Z.; Mo, F.; Tang, Z.; Li, H.; Wang, Z.; Sarangi, V.; Pramanick, A.; Fan, J.; Zhi, C. Self-healable electroluminescent devices. *Light. Sci. Appl.* **2018**, *7*, 1–11. [CrossRef] [PubMed]
12. Park, H.J.; Kim, S.; Lee, J.H.; Kim, H.T.; Seung, W.; Son, Y.; Kim, T.Y.; Khan, U.; Park, N.M.; Kim, S.W. Self-powered motion-driven triboelectric electroluminescence textile system. *ACS Appl. Mater. Interfaces* **2019**, *11*, 5200–5207. [CrossRef]
13. Shin, Y.B.; Ju, Y.H.; Lee, H.J.; Han, C.J.; Lee, C.R.; Kim, Y.; Kim, J.W. Self-Integratable, Healable and Stretchable Electroluminescent Device Fabricated via Dynamic Urea Bonds Equipped in Polyurethane. *ACS Appl. Mater. Interfaces* **2020**, *12*, 10949–10958. [CrossRef]
14. Lee, S.; Kang, T.; Lee, W.; Afandi, M.M.; Ryu, J.; Kim, J. Multifunctional device based on phosphor-piezoelectric PZT: Lighting, speaking, and mechanical energy harvesting. *Sci. Rep.* **2018**, *8*, 1–7. [CrossRef]
15. Kim, J.S.; Cho, S.H.; Kim, K.L.; Kim, G.; Lee, S.W.; Kim, E.H.; Jeong, B.; Hwang, I.; Han, H.; Shim, W.; et al. Flexible artificial synesthesia electronics with sound-synchronized electroluminescence. *Nano Energy* **2019**, *59*, 773–783. [CrossRef]

16. Cho, S.; Kang, D.H.; Lee, H.; Kim, M.P.; Kang, S.; Shanker, R.; Ko, H. Highly Stretchable Sound-in-Display Electronics Based on Strain-Insensitive Metallic Nanonetworks. *Adv. Sci.* **2020**, *8*, 2001647. [CrossRef] [PubMed]
17. Sandanayaka, A.S.D.; Matsushima, T.; Bencheikh, F.; Terakawa, S.; Potscavage, W.J.P., Jr.; Qin, C.; Fujihara, T.; Goushi, K.; Ribierre, J.-C.; Adachi, C. Indication of current-injection lasing from an organic semiconductor. *Appl. Phys. Express* **2019**, *12*, 061010. [CrossRef]
18. Moshaev, V.; Leibin, Y.; Malka, D. Optimizations of Si PIN diode phase-shifter for controlling MZM quadrature bias point using SOI rib waveguide technology. *Opt. Laser Technol.* **2021**, *138*, 106844. [CrossRef]
19. Malka, D. A Four Green TM/Red TE Demultiplexer Based on Multi Slot-Waveguide Structures. *Materials* **2020**, *13*, 3219. [CrossRef] [PubMed]
20. Nabesaka, K.; Ishikawa, Y.; Hosokawa, Y.; Uraoka, Y. Photomechanical modification of ZnS microcrystal to enhance electroluminescence by ultrashort-pulse laser processing. *Appl. Phys. Express.* **2017**, *10*, 021201. [CrossRef]
21. Tan, Y.J.; Godaba, H.; Chen, G.; Tan, S.T.M.; Wan, G.; Li, G.; Lee, P.M.; Cai, Y.; Li, S.; Shepherd, R.F.; et al. A transparent, self-healing and high-κ dielectric for low-field-emission stretchable optoelectronics. *Nat. Mater.* **2020**, *19*, 182–188. [CrossRef] [PubMed]
22. Satoh, T.; Nakatsuta, T.; Tsuruya, K.; Tabata, Y.; Tamura, T.; Ichikawa, Y.; Tango, H. Electrical properties of two-sided luminescence powder-distributed inorganic electroluminescence panels. *J. Mater. Sci. Mater. Electron.* **2007**, *18*, 239–242. [CrossRef]
23. Kim, J.-Y.; Bae, M.J.; Park, S.H.; Jeong, T.; Song, S.; Lee, J.; Han, I.; Yoo, J.B.; Jung, D.; Yu, S. Electroluminescence enhancement of the phosphor dispersed in a polymer matrix using the tandem structure. *Org. Electron.* **2011**, *12*, 529–533. [CrossRef]
24. Takeda, N.; Tsuneyasu, S.; Satoh, T. EL emission enhancement in powder electroluminescent device via insertion of receptive layer. *Electron. Lett.* **2020**, *56*, 147. [CrossRef]
25. Kim, J.-Y.; Park, S.H.; Jeong, T.; Bae, M.J.; Song, S.; Lee, J.; Han, I.T.; Jung, D.; Yu, S. Paper as a Substrate for Inorganic Powder Electroluminescence Devices. *IEEE Trans. Electron Devices* **2010**, *57*, 1470–1474. [CrossRef]
26. Sloma, M.; Janczak, D.; Wroblewski, G.; Mlozniak, A.; Jakubowska, M. Electroluminescent structures printed on paper and textile elastic substrates. *Circuit World* **2014**, *40*, 13–16. [CrossRef]
27. Yakoh, A.; Siangproh, W.; Chailapakul, O.; Ngamrojanavanich, N. Optical Bioelectronic Device Based on a Screen-Printed Electroluminescent Transducer. *ACS Appl. Mater. Interfaces* **2020**, *12*, 22543–22551. [CrossRef] [PubMed]
28. Fukuzumi, H.; Saito, T.; Iwata, T.; Kumamoto, Y.; Isogai, A. Transparent and High Gas Barrier Films of Cellulose Nanofibers Prepared by TEMPO-Mediated Oxidation. *Biomacromolecules* **2009**, *10*, 162–165. [CrossRef]
29. Nogi, M.; Iwamoto, S.; Nakagaito, A.N.; Yano, H. Optically Transparent Nanofiber Paper. *Adv. Mater.* **2009**, *21*, 1595–1598. [CrossRef]
30. Nogi, M.; Kim, C.; Sugahara, T.; Inui, T.; Takahashi, T.; Suganuma, K. High thermal stability of optical transparency in cellulose nanofiber paper. *Appl. Phys. Lett.* **2013**, *102*, 181911. [CrossRef]
31. Rodionova, G.; Eriksen, Ø.; Gregersen, Ø. TEMPO-oxidized cellulose nanofiber films: Effect of surface morphology on water resistance. *Cellulose* **2012**, *19*, 1115–1123. [CrossRef]
32. Nakagaito, A.N.; Iwamoto, S.; Yano, H. Bacterial cellulose: The ultimate nano-scalar cellulose morphology for the production of high-strength composites. *Appl. Phys. A Mater. Sci. Process.* **2005**, *80*, 93–97. [CrossRef]
33. Henriksson, M.; Berglund, L.A.; Isaksson, P.; Lindström, T.; Nishino, T. Cellulose nanopaper structures of high toughness. *Biomacromolecules* **2008**, *9*, 1579–1585. [CrossRef]
34. Nogi, M.; Abe, K.; Handa, K.; Nakatsubo, F.; Ifuku, S.; Yano, H. Property enhancement of optically transparent bionanofiber composites by acetylation. *Appl. Phys. Lett.* **2006**, *89*, 233123. [CrossRef]
35. Uetani, K.; Okada, T.; Oyama, H.T. In-Plane Anisotropic Thermally Conductive Nanopapers by Drawing Bacterial Cellulose Hydrogels. *ACS Macro Lett.* **2017**, *6*, 345–349. [CrossRef]
36. Wise, L.E. Chlorite holocellulose, its fractionation and bearing in summative wood analysis and on studies in hemicellulose. *Pap. Trade J.* **1946**, *122*, 35–43.
37. Isogai, A.; Saito, T.; Fukuzumi, H. TEMPO-oxidized cellulose nanofibers. *Nanoscale* **2011**, *3*, 71–85. [CrossRef]
38. Okita, Y.; Saito, T.; Isogai, A. Entire Surface Oxidation of Various Cellulose Microfibrils by TEMPO-Mediated Oxidation. *Biomacromolecules* **2010**, *11*, 1696–1700. [CrossRef] [PubMed]
39. Saito, T.; Isogai, A. TEMPO-Mediated Oxidation of Native Cellulose. The Effect of Oxidation Conditions on Chemical and Crystal Structures of the Water-Insoluble Fractions. *Biomacromolecules* **2004**, *5*, 1983–1989. [CrossRef]
40. Uetani, K.; Okada, T.; Oyama, H.T. Crystallite Size Effect on Thermal Conductive Properties of Nonwoven Nanocellulose Sheets. *Biomacromolecules* **2015**, *16*, 2220–2227. [CrossRef]
41. Uetani, K.; Hatori, K. Thermal conductivity analysis and applications of nanocellulose materials. *Sci. Technol. Adv. Mater.* **2017**, *18*, 877–892. [CrossRef] [PubMed]
42. Wang, S.G.; Zhang, Q.; Yoon, S.F.; Ahn, J.; Wang, Q.; Yang, D.J.; Zhou, Q.; Yue, N.L. Optical properties of nano-crystalline diamond films deposited by MPECVD. *Opt. Mater.* **2003**, *24*, 509–514. [CrossRef]
43. Ji, J.; Perepichka, I.F.; Bai, J.; Hu, D.; Xu, X.; Liu, M.; Wang, T.; Zhao, C.; Meng, H.; Huang, W. Three-phase electric power driven electroluminescent devices. *Nat. Commun.* **2021**, *12*, 1–11.
44. Kim, J.Y.; Jeong, T.; Park, S.H.; Bae, M.J.; Song, S.; Lee, J.; Han, I.; Jung, D.; Liu, C.; Yu, S.G. Enhanced optical and electrical properties of inorganic electroluminescent devices using the top-emission structure. *Jpn. J. Appl. Phys.* **2011**, *50*, 06GF06. [CrossRef]

 *nanomaterials*

Article

Enzymatic Preparation and Characterization of Spherical Microparticles Composed of Artificial Lignin and TEMPO-Oxidized Cellulose Nanofiber

Naoya Fukuda, Mayumi Hatakeyama and Takuya Kitaoka *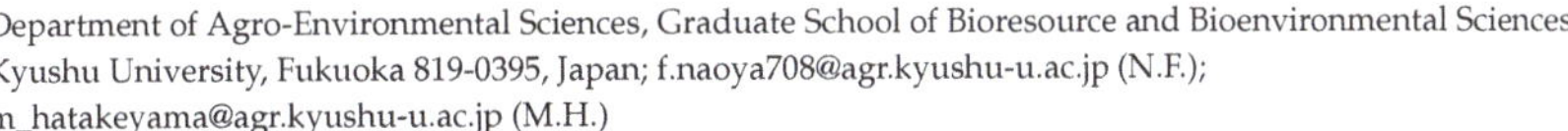

Department of Agro-Environmental Sciences, Graduate School of Bioresource and Bioenvironmental Sciences, Kyushu University, Fukuoka 819-0395, Japan; f.naoya708@agr.kyushu-u.ac.jp (N.F.); m_hatakeyama@agr.kyushu-u.ac.jp (M.H.)
* Correspondence: tkitaoka@agr.kyushu-u.ac.jp

Abstract: A one-pot and one-step enzymatic synthesis of submicron-order spherical microparticles composed of dehydrogenative polymers (DHPs) of coniferyl alcohol as a typical lignin precursor and TEMPO-oxidized cellulose nanofibers (TOCNFs) was investigated. Horseradish peroxidase enzymatically catalyzed the radical coupling of coniferyl alcohol in an aqueous suspension of TOCNFs, resulting in the formation of spherical microparticles with a diameter and sphericity index of approximately 0.8 μm and 0.95, respectively. The ζ-potential of TOCNF-functionalized DHP microspheres was about −40 mV, indicating that the colloidal systems had good stability. Nanofibrous components were clearly observed on the microparticle surface by scanning electron microscopy, while some TOCNFs were confirmed to be inside the microparticles by confocal laser scanning microscopy with Calcofluor white staining. As both cellulose and lignin are natural polymers known to biodegrade, even in the sea, these woody TOCNF−DHP microparticle nanocomposites were expected to be promising alternatives to fossil resource-derived microbeads in cosmetic applications.

Keywords: dehydrogenative polymer; enzymatic radical coupling; lignin; nanocellulose; microsphere; TEMPO-oxidized cellulose nanofiber

Citation: Fukuda, N.; Hatakeyama, M.; Kitaoka, T. Enzymatic Preparation and Characterization of Spherical Microparticles Composed of Artificial Lignin and TEMPO-Oxidized Cellulose Nanofiber. *Nanomaterials* **2021**, *11*, 917. https://doi.org/10.3390/nano11040917

Academic Editor: Linda J. Johnston

Received: 13 March 2021
Accepted: 2 April 2021
Published: 3 April 2021

Publisher's Note: MDPI stays neutral with regard to jurisdictional claims in published maps and institutional affiliations.

1. Introduction

The adverse impact of microplastics on marine ecosystems has become an urgent issue to solve [1–5]. In general, microplastics are small plastic debris derived from various nonbiodegradable polymer materials used in daily life [3]. Initially, the gradual downsizing of large polymer materials, such as plastic bags and bottles, to small fractions, known as secondary microplastics, was considered a major issue. However, recently, primary microplastics, such as microparticles (MPs) contained in facial cleaners and cosmetics, have resulted in more serious problems [4,5]. These microplastics flow out to the ocean and accumulate in the sea for a long time without any significant biodegradation. Therefore, primary small plastics must immediately be substituted with biodegradable alternatives, with many studies on enhancing the biodegradability of synthetic polymers having been conducted to develop environmentally friendly polymer microparticles [6–9].

Cellulose and lignin are two major components of wood cell walls [10,11]. Cellulose is a linear homopolymer composed only of β-D-glucopyranose, which forms crystalline microfibrils during biosynthesis owing to regular strong inter- and intra-molecular hydrogen bonds. The microfibrils in wood greatly contribute to its tough structure, endowing trees with rigidity [12–14]. Cellulose is well known as biodegradable, owing to enzymatic decomposition mediated by *endo-/exo*-cellulases and β-glucosidases secreted by fungi, bacteria, and microorganisms found in forests and the deep sea [15–17]. Meanwhile, lignin is an aromatic polymer formed via enzymatic radical coupling of several tautomeric isomers of phenylpropanoid monolignols [18]. Lignin in wood absorbs ultraviolet light and endows the wood architecture with hydrophobicity through accumulation between the cell walls, known as lignification [18,19]. Lignin is also enzymatically degraded by the action of white-rot fungi [20,21]. Furthermore, recent research has shown that deep-sea microorganisms, such as *Novosphingobium* strains, can break down lignin-specific structures [22]. Therefore, many studies have focused on developing sustainable, biodegradable, and ecofriendly materials from woody biomass for various applications [14,23–25].

In the last two decades, a new type of cellulose nanomaterial, cellulose nanofiber (CNF), has attracted much attention owing to its extraordinary nanoarchitectures and inherent crystalline fiber structure [14,23,24,26]. This novel nanomaterial is expected to have applications as a natural filler for reinforced plastics [27–29], food and cosmetic additive [30,31], flexible electronic devices [32,33], and in optoelectronics [34,35] owing to its fascinating physicochemical properties [14,23]. Recently, various MPs prepared by CNF-stabilized Pickering emulsion templating methods, in which thin CNF shells cover polymer cores, have received much interest regarding practical applications [36–39]. These CNF-integrated MPs have shown high pH-responsive properties [36], adsorption/desorption abilities for drug loading/release [37], and temperature-regulation capabilities [38]. Furthermore, CNF-stabilized MPs are expected to have cosmetic and medical applications. Fujisawa et al. reported using suspension polymerization to synthesize polystyrene and poly(divinylbenzene) MPs from corresponding monomers in an oil phase, stabilized by 2,2,6,6-tetramethylpiperidine 1-oxyl (TEMPO)-oxidized CNFs (TOCNFs) [36,37,39]. These TOCNF-covered MPs have unique core–shell structures, and TOCNFs decompose easily [16]. However, the polymer cores in the obtained MPs are not biodegradable, potentially resulting in marine pollution [1,5]. Therefore, biodegradable components are promising for use in the synthesis of MPs for cosmetic applications. In this context, cellulose and lignin are attractive natural resources expected to undergo microbial degradation in the deep sea. Therefore, the preparation of biodegradable MPs by combining CNF and lignin is a promising approach to the functional design of marine-degradable MPs.

In previous reports, lignin derivatives extracted from natural woods by various procedures have been used to fabricate MPs [40,41]. However, nature-derived lignin has crude and complex structures, which strongly affect its biodegradability and biocompatibility. On the other hand, to determine the biosynthesis mechanism of lignin in vivo, the in vitro synthesis of dehydrogenative polymers (DHPs) of coniferyl alcohol has long been investigated for use as an artificial lignin, enabling rough control of molecular structures [29,42,43]. Mićić et al. found that synthetic DHP molecules self-assemble to form fine particles [44,45]. Although several researchers have attempted to prepare MP composites using cellulose materials, few attempts to design spherical MPs combined with CNF have been reported because cellulose has a low affinity for DHPs, resulting in phase separation to form large aggregates [46,47]. Recently, our previous study revealed the affinity of CNF for DHP in the interfacial enzymatic synthesis of CNF–DHP nanocomposites [29]. In this article, we report the preliminary results of in situ enzymatic synthesis of DHPs from coniferyl alcohol in the presence of TOCNF without any organic solvents. TEMPO-based biomass conversion is expected as one of the promising green approaches to produce various nanomaterials such as CNFs isolated by TEMPO oxidation [48], chemo-enzymatically nanofibrillated lignocellulose [49], and biomass-derived chemicals by TEMPO-mediated catalysis [50]. In this work, nanoscale reconstruction of woody components was investigated to form cosmetic microspheres by using TOCNFs and DHPs, respectively, as surface-carboxylated cellulose microfibrils and artificial lignin (Figure 1). This one-pot green synthesis of spherical submicron-order TOCNF–DHP composite MPs is expected to provide new insight into the fabrication of marine-degradable microplastics.

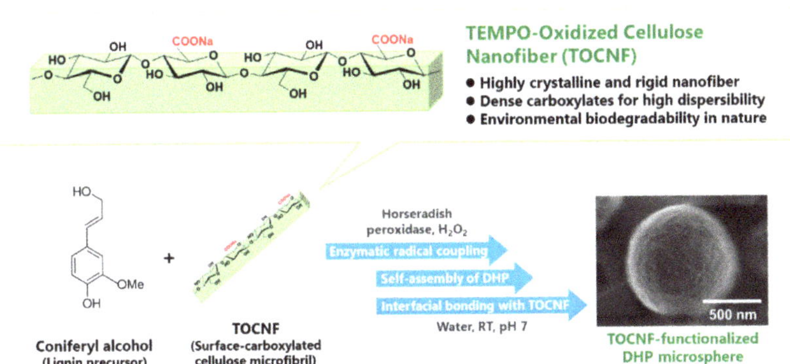

Figure 1. Schematic illustration of the research strategy: a one-pot and one-step green synthesis of spherical submicron-order TEMPO-oxidized cellulose nanofiber (TOCNF)-functionalized dehydrogenative polymer (DHP) microparticles.

2. Materials and Methods

2.1. Materials

TEMPO-oxidized CNF (TOCNF) used in this study was kindly provided by DKS Co. Ltd., Kyoto, Japan (RHEOCRYSTA, I-2SX, 2.3% (w/v), COONa = 1.55 mmol g^{-1} of TOCNF). Coniferyl alcohol, hydrogen peroxide (H$_2$O$_2$, 30% aqueous solution), and horseradish peroxidase (HRP) were purchased from Wako Pure Chemical Industries Ltd. (Osaka, Japan). Other chemicals and solvents were purchased from Sigma-Aldrich Japan, Ltd. (Tokyo, Japan), Wako Pure Chemical Industries, Ltd. (Osaka, Japan), and Tokyo Chemical Industry Co., Ltd. (Tokyo, Japan). All chemicals were used as received without further purification. Water used in this study was purified using a Barnstead Smart2Pure system (Thermo Scientific Co. Ltd., Tokyo, Japan).

2.2. Characterization of TOCNF

The nano-order morphology of TOCNF was observed using transmission electron microscopy (TEM; JEM-2100HCKM, JEOL Ltd., Tokyo, Japan) by staining with sodium phosphotungstate, and scanning probe microscopy (SPM; Dimension Icon, Bruker Japan Co. Ltd., Tokyo, Japan) equipped with a SCANASYST-AIR probe (k = 0.4 N m^{-1}, F_0 = 70 kHz) (Figure 2a,b, respectively). The crystalline structure of TOCNF was recorded by X-ray diffraction (XRD) analysis using a Rigaku SmartLab diffractometer (Rigaku Co., Tokyo, Japan) with Ni-filtered Cu Kα radiation (λ = 0.15418 nm) at 40 kV and 20 mA. The scanning rate was 0.5° min^{-1} with 0.05° intervals (Figure 2c). The carboxylate content of the TOCNF was determined by electrical conductivity titration [48]. The obtained data for TOCNF corresponded well with reported data [48]. TEM was performed at the Ultramicroscopy Research Center, Kyushu University. SPM and XRD analyses were conducted at the Center of Advanced Instrumental Analysis, Kyushu University.

2.3. Preparation of TOCNF-Functionalized DHP Microparticles

TOCNF–DHP MPs were enzymatically synthesized using the conventional DHP synthesis protocol, known as the Zulaufverfahren (ZL) method, i.e., "bulk" polymerization method [29,42]. Briefly, coniferyl alcohol (50 mg), 30% H$_2$O$_2$ (50 µL), and pure water (950 µL) were added to TOCNF suspension (0–1.0% (w/v), 4 mL). HRP solution (500 µL, 500 µg mL^{-1} in pure water) was then added to the reaction system. The reaction mixture was gently stirred at room temperature for 24 h to obtain TOCNF–DHP MPs, which were collected by centrifugation and thoroughly washed with purified water. The MP recovery ratios were up to 85% by weight. Another conventional DHP synthesis method, Zutropfverfahren (ZT; end-wise polymerization method) [29,42] was conducted as a control. The as-prepared MPs were observed using an optical microscope (DMI 4000B, Leica,

Wetzlar, Germany) by mounting a droplet of the MP suspension on a glass slide. The ζ-potential of TOCNF–DHP MPs in an aqueous medium was measured at pH 7.0 using a Zetasizer Nano ZS instrument equipped with a 632-nm HeNe laser operating at a 173-degree detector angle (Malvern Panalytical, Tokyo, Japan). Fourier transform infrared (FTIR) spectroscopy was carried out to characterize as-prepared MPs using an FT/IR-620 spectrometer (JASCO, Tokyo, Japan). Each freeze-dried sample (ca. 2 mg) was mixed with 200 mg of KBr, followed by pressing it into transparent pellets. These samples were analyzed in the spectral range of 400–4000 cm^{-1} with a resolution of 2 cm^{-1}.

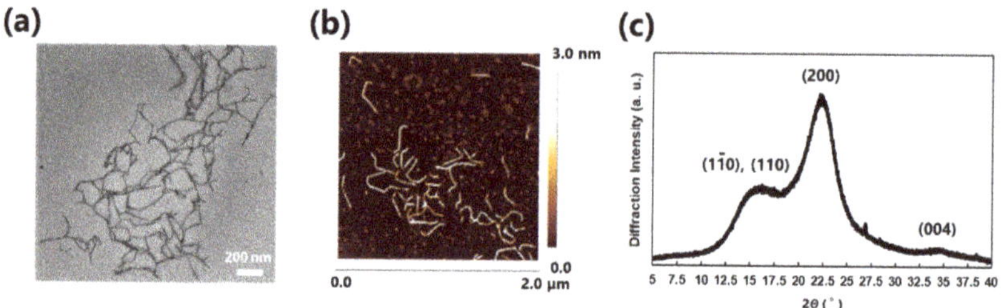

Figure 2. Characterization data of TOCNF used in this study. (**a**) transmission electron microscopy (TEM) image, (**b**) scanning probe microscopy (SPM) image (average height of each fiber = 2.1 ± 0.7 nm, n = 60 for each sample), and (**c**) X-ray diffraction (XRD) profile showing a typical cellulose I crystalline structure.

2.4. Scanning Electron Microscopy (SEM)

MP suspension (1.0% (w/v), 3 μL) was dropped on a carbon tape and dried in a desiccator at room temperature for 24 h. These samples were then coated using an osmium coater (HPC-1SW, Vacuum Device, Ibaraki, Japan) for 1 s. SEM (SU8000, Hitachi High-Tech Co, Tokyo, Japan) analysis was performed at 0.7–1.0 kV. The diameter and sphericity index of MPs were manually calculated from the digital SEM images obtained. In brief, the orthogonal minor axis was divided by the major axis for 80 particles of each sample. The experiment was repeated twice to confirm the reproducibility. The cross-section of MP samples embedded in resin, cut using an ultramicrotome (Ultracut-UCT, Leica, Wetzlar, Germany), was observed by SEM analysis, performed at the Center of Advanced Instrumental Analysis, Kyushu University.

2.5. Confocal Laser Microscopy (CLSM)

MP suspension (1.0% (w/v), 2 μL) was mixed with 0.001% Calcofluor white stain solution (2 μL) on a glass slide, and the sample was then dried in the dark at room temperature for 24 h. The MPs were observed by confocal laser microscopy (TCS SP8 STED, Leica, Wetzlar, Germany) with excitation/emission at 405/415–500 nm, respectively. CLSM analysis was conducted at the Center for Advanced Instrumental and Educational Supports, Faculty of Agriculture, Kyushu University.

2.6. Drug Loading Test

Methylene blue (MB) and arginine (Arg) were used as model drugs for loading on MPs. MB and arginine hydrochloride were dissolved in phosphate buffer (pH 7.0, 10 mM) and sodium hydrogen carbonate aqueous solution (pH 8.4, 50 mM), respectively. The initial concentration of each model drug was adjusted to 5.0 mg L^{-1}. Freeze-dried MPs (10 mg) were dispersed in the solution (1.0 mL) and stirred at room temperature for 2 h. The MP suspension was filtered through a 0.20-μm polytetrafluoroethylene membrane filter. The concentration of MB in the filtrate was determined using a UV–vis spectrophotometer (UH5300, Hitachi High-Tech Co., Tokyo, Japan) at the absorbance of MB (λ = 665 nm). The final Arg concentration in the solute was determined by reversed-phase high-performance liquid chromatography (RP-HPLC) analysis (Prominence UFLC, Shimadzu Co., Kyoto, Japan), by determining the peak intensity of dabsylated Arg [51]. Separation was achieved using a Shim-pack XR-ODS II column. Gradient elution mode with a flow rate of 1.0 mL min^{-1} was used, with 0.05% phosphate solution and acetonitrile used as mobile phases. The UV detector wavelength was set to 450 nm.

3. Results and Discussion

3.1. Preparation of Wood-Mimetic CNF–Lignin MPs

DHPs were synthesized from coniferyl alcohol by HRP-mediated catalysis, namely the ZL method, in which all reactants are mixed together in one pot to generate radicals rapidly in 0.8% (w/v) TOCNF suspension. After washing MPs with pure water to remove excess TOCNF, spherical MPs were observed by optical microscopy (Figure 3a) [44,45]. In the absence of either HRP or H_2O_2, no MP formation occurred, as shown in Figure 3b,c, respectively. These results indicated that the MPs were enzymatically synthesized, and presumably composed of polymerized DHPs, but not precipitates of coniferyl alcohol monomer [29,42]. DHP MPs were also synthesized by the stepwise Zutropfverfahren (ZT) method, in which the radical generation rate is relatively slow [46]. ZT-type DHPs synthesized with TOCNF afforded irregular-shaped aggregates in addition to spherical MPs (Figure 3d). Sipponen et al. reported the successful fabrication of lignin nanoparticles by nanoprecipitation from aqueous ethanol, where initial precipitation of lignin formed small nuclei, followed by adsorption of other lignin components onto the as-formed nuclei, which resulted in nanoparticle growth [52]. When lignin precipitation occurs rapidly, numerous small nuclei are formed, eventually hindering the further growth of nanoparticles. In our study, the one-pot ZL method possibly promoted the formation of many small spherical precipitates in the presence of TOCNF, while slow generation of DHP precipitates by the ZT method caused gradual DHP deposition, resulting in the formation of relatively large nanoparticles. Furthermore, in the synthesis of ZT-type DHPs with TOCNF, coexisting TOCNF might disturb the smooth generation of DHPs, resulting in irregular-shaped aggregates. In contrast, the ZL method allowed the fabrication of spherical MPs through rapid radical formation and then self-assembly of the as-prepared DHPs, without inhibition by TOCNF. Morphological characterization data from SEM and CLSM analyses are presented in detail in the following section.

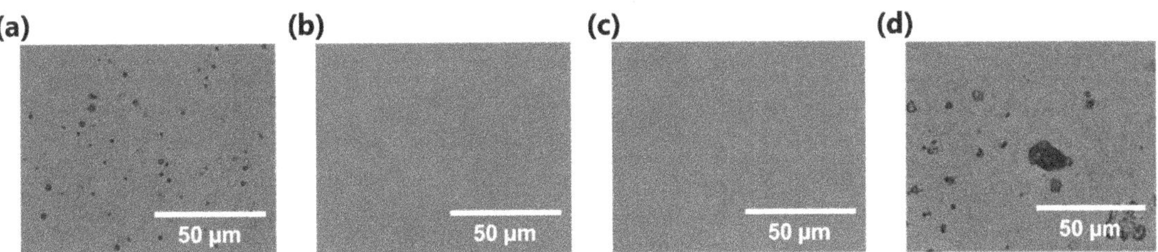

Figure 3. Optical images of DHP microparticles (MPs) synthesized (**a**) in the presence of 0.8% (w/v) TOCNF with HRP and H_2O_2, (**b**) w/o HRP, and (**c**) w/o H_2O_2 using the Zulaufverfahren (ZL) method. (**d**) Zutropfverfahren (ZT) method using 0.8% (w/v) TOCNF with HRP and H_2O_2.

3.2. Morphological Characterization of TOCNF–DHP MPs

Optical microscopy images of MPs prepared using various concentrations of TOCNF are shown in Figure 4a–e. Spherical MPs were successfully synthesized with 0–0.8% (w/v) TOCNF suspension. Meanwhile, an irregular shape was observed in the case of 1.0% (w/v) TOCNF, owing to the higher TOCNF concentration affording a viscous suspension that inhibited DHP self-assembly [53]. The nanomorphology of as-prepared MPs was further investigated by SEM analysis (Figure 4f–j). As with optical microscopy results, TOCNF concentrations below 0.8% (w/v) resulted in formation of spherical MPs, possibly due to smooth DHP self-assembly. An increase in the TOCNF concentration provided MPs with fewer bumpy surfaces. Furthermore, nanofibrous components were observed on the MP surface, presumably indicating TOCNF fractions present on MP surfaces (inset of Figure 4i) [36,37], although TOCNF-free MP did not exhibit any similar components on the surface in the inset of Figure 4f. Therefore, our one-step green synthesis of wood-mimetic MPs enabled the formation of TOCNF-covered MPs containing DHP cores.

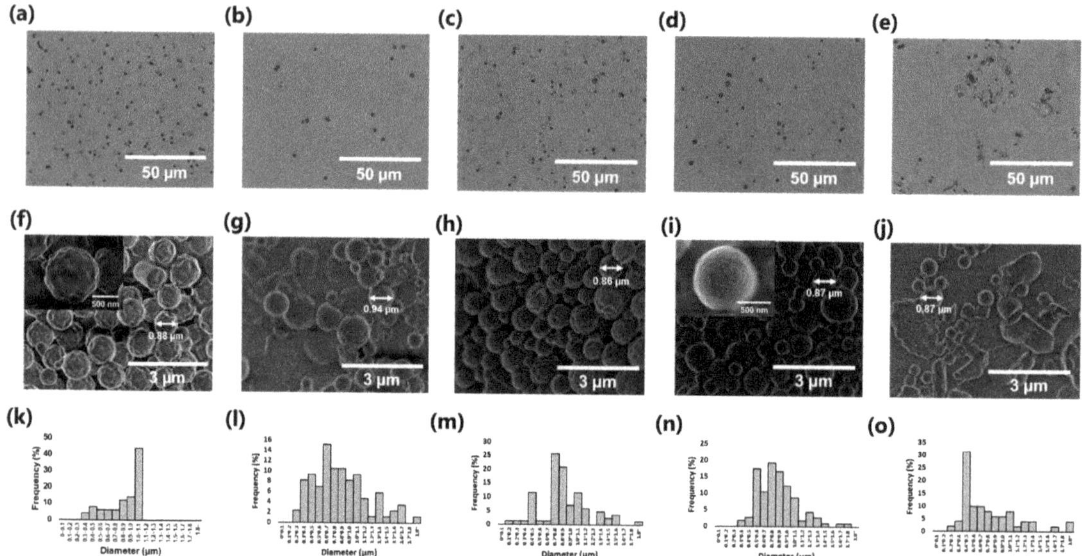

Figure 4. (**a–e**) Optical images, (**f–j**) SEM images, and (**k–o**) particle size distribution histograms of MPs synthesized with different concentrations of TOCNF: (**a,f,k**) 0% (w/v), (**b,g,l**) 0.3% (w/v), (**c,h,m**) 0.5% (w/v), (**d,i,n**) 0.8% (w/v), and (**e,j,o**) 1.0% (w/v). Insets of (**f,i**) show magnified picture of each one microparticle.

Spectroscopic characterization by FTIR analysis clearly revealed the successful formation of typical guaiacyl (G)-type DHP (Figure 5) [54]. A characteristic carbonyl peak at 1610 cm^{-1}, originating from TOCNF [48], overlapped with the C=C stretching peak of G-rings at 1601 cm^{-1}; however, the relative peak intensity at 1601/1510 for G-ring stretching of DHP increased from 0.487 to 0.544 with increasing amount of TOCNF, ranging from 0 to 0.5% (w/v). Therefore, as-prepared MPs were considered to contain the TOCNF components. Excess amount of TOCNF reduced the peak intensity around 1600 cm^{-1}, possibly indicating the MP formation in an inappropriate fashion. Neither X-ray diffraction analysis nor X-ray photoelectron spectroscopy provided any significant peaks (data not shown), due to a slight amount of TOCNF on the MP surface below each detection limit.

The particle diameter and sphericity index of MPs were calculated from the SEM images (Figure 4f–i). The average diameter of MPs was about 0.8 μm, in the submicrometer (submicron) order, as listed in Table 1, which is expected to have practical cosmetic applications [55]. Furthermore, the standard deviation of each diameter slightly decreased as the TOCNF content increased from 0.3% (w/v) to 0.8% (w/v), while the diameter was unchanged. This was possibly due to TOCNF adsorbed on the MP surface acting like a Pickering emulsion stabilizer to improve MP stability in water [56]. The MP sphericity index was up to 0.94. True spheres are an important requirement of cosmetic use to improve cream spreadability of the foundation matrix [57]. The MP surface charges were determined by ζ-potential measurement (Table 1), showing −32.5 mV for CNF-free DHPs, which was consistent with the results of natural lignin nanoparticles [58]. The ζ-potential values of TOCNF–DHP MPs were up to about −40.0 mV, indicating negatively charged TOCNF adsorbed on MPs, which is expected to improve stability of the particle dispersion [59]. It has been reported that wood-derived TOCNFs have strong negative charges in the pH range of 5 to 7, equal to or greater than that at pH 7 [60]; therefore, as-prepared TOCNF–DHP MPs are expected to effectively use in a slightly acidic environment on the surfaces of human skins. Be that as it may, a long-term stability test is required to evaluate product applicability at practical levels.

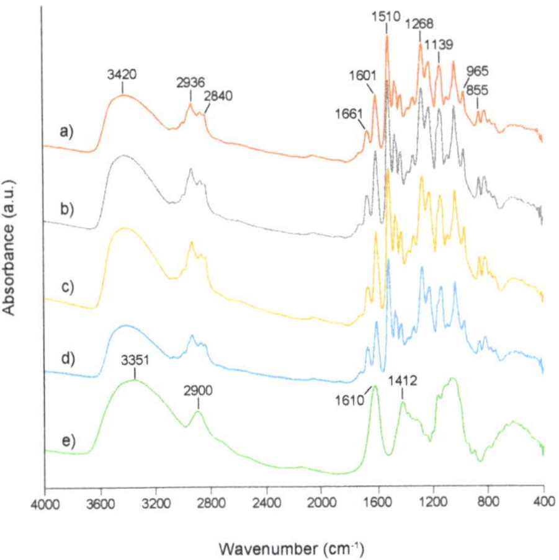

Figure 5. FTIR spectra of MPs synthesized with different concentrations of TOCNF: (**a**) 0% (*w/v*), (**b**) 0.3% (*w/v*), (**c**) 0.5% (*w/v*), (**d**) 0.8% (*w/v*), and (**e**) TOCNF used.

Table 1. Diameter, sphericity index, and ζ-potential of TOCNF–DHP MPs.

TOCNF Concentration (% (*w/v*))	Diameter (µm) [1]	Sphericity [1]	ζ-Potential (mV) [2]
0.0	0.85 ± 0.24	0.97 ± 0.03	-32.5 ± 6.8
0.3	0.83 ± 0.36	0.95 ± 0.08	-39.8 ± 7.9
0.5	0.88 ± 0.31	0.94 ± 0.08	-40.7 ± 7.1
0.8	0.83 ± 0.26	0.95 ± 0.07	-38.4 ± 7.2
1.0	0.77 ± 0.49	0.80 ± 0.21	-40.4 ± 7.8

[1] Calculated from SEM images at $n = 80$ for each sample. [2] Measured at $n = 5$ for each sample.

Purified TOCNF–DHP MPs were freeze-dried to obtain solid MPs in powder form. In general, natural lignin is a dark brown color, which causes practical problems in cosmetic use [61]. However, the DHP MP powder had a creamy white color, as shown in the insets of Figure 6a–d. Very fine MPs were observed by SEM analysis (Figure 6a–d), but a high TOCNF concentration of 0.8% (*w/v*) seemed to induce morphological collapse of the MPs. This might be due to excess TOCNF on the MP surface aggregating during the freeze-drying process, resulting in some MP distortion. Herein, TOCNF–DHP MPs synthesized with 0.5% (*w/v*) TOCNF showed good shapes and were used in the following tests.

To evaluate the internal structure of the MPs, the MP cross-section was observed by SEM analysis. The sample embedded in resin was sliced using an ultramicrotome for SEM observation. The sliced image was not clear, but roughly exhibited furry TOCNF surrounding MPs, as shown in Figure 6e. This implied that TOCNF–DHP MPs had a shell layer of TOCNF [37]. CLSM analysis using Calcofluor white dye was conducted to visualize TOCNF inside the MPs (Figure 6f–i) [56]. TOCNF–DHP MPs showed strong fluorescence derived from Calcofluor white adsorbed to TOCNF (Figure 6h). In contrast, no significant fluorescence was observed in DHP MPs prepared without TOCNF (Figure 6f). These results indicated that fluorescent dye molecules selectively adsorbed to TOCNF. Furthermore,

Z-stack imaging was conducted to show the interior distribution of TOCNF (Figure 6j). Each sliced image of the MPs was taken at several depth levels, ranging from the top to the middle of the MPs. In all pictures, fluorescence was detected over the whole MP cross-section. This suggested that TOCNF was embedded inside the MPs as well as on the surface. When DHPs were synthesized in the presence of TOCNF using the ZL method, some TOCNF was captured in the spherical MPs formed, possibly through hydrogen bonding and non-covalent π–interactions [62]. The TOCNFs adsorbed on the MP surface were stabilized by van der Waals interaction [63]. Covalent bond formation like lignin–carbohydrates complexes (LCCs) may be possible [64]. Accordingly, most TOCNF on the MPs formed a shell structure, while a significant amount of TOCNF was also embedded inside the MPs, although TOCNF inside the MPs was not observed in the SEM image (Figure 5e). CNF has high crystallinity and high mechanical strength, and has been widely used as a filler for reinforced plastics [27,28]. Therefore, incorporated TOCNF, exhibiting a rigid nanofiber form with the natural crystalline structure of cellulose I, as shown in Figure 2, would contribute to reinforcing the MPs when synthesized with TOCNF concentration of up to 0.5% (w/v). Moreover, different types of nanocellulose such as cellulose nanocrystals (CNCs) are also expected to use as a promising candidate for producing spherical CNC–DHP MPs, due to higher rigidity with more ordered alignments, possibly causing the better surface coverage and physical properties of MPs.

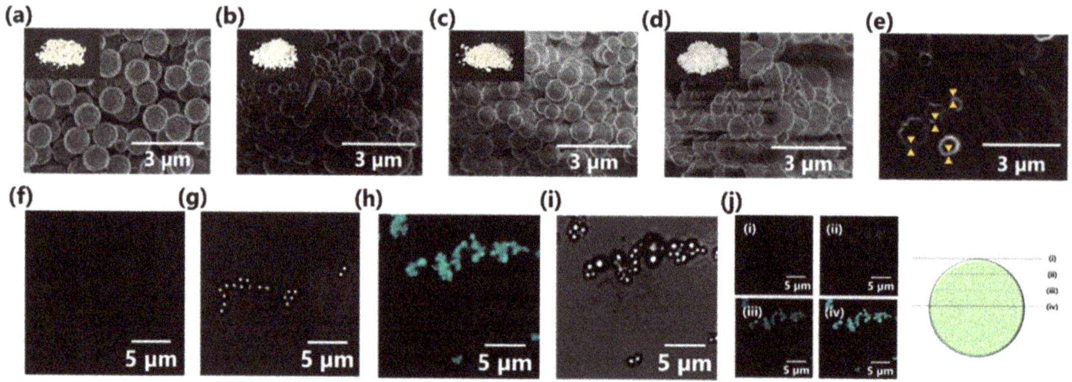

Figure 6. Morphology of freeze-dried MP powders prepared using different TOCNF concentrations: (**a**) 0% (w/v), (**b**) 0.3% (w/v), (**c**) 0.5% (w/v), and (**d**) 0.8% (w/v). Each inset corresponds to the optical image. (**e**) Cross-section SEM image of MPs from (**c**), with orange triangle markers indicating TOCNF-shell structures. CLSM images of MPs stained with Calcofluor white: (**f,g**) DHP MPs and (**h**−**j**) TOCNF–DHP MPs (TOCNF concentration, 0.5% (w/v)). (**f,h,j**) are fluorescent images. (**g,i**) are bright field images. Each panel of (**j**) is a Z-stacking image at each position, as illustrated.

3.3. Preliminary Test for Drug-Loading on MPs

Cosmetic MPs require many practical performances such as long-term stability [65], biocompatibility [66], and UV-absorbing property [67]. Besides, drug-loading capability is one of the important factor for MP use in cosmetic applications [68]. In this study, two model compounds were subjected to adsorption behavior tests, namely, methylene blue (MB, a major model drug [37,69]) and arginine (Arg, a typical moisturizing ingredient for human skin [70]). MPs were poured into water solution containing each model drug, and stirred for 2 h. The MPs maintained their original spherical structures without any TOCNF peeling from the MP surface (Figure 7a–d). The drug loading was determined to measure the residual amount of each component in the supernatant. MB and Arg loadings of 94.9 ± 1.8% and 38.3 ± 0.2%, respectively, were obtained on the TOCNF–DHP MPs. These results indicated that TOCNF–DHP MPs had significant drug-loading capabilities. However, this preliminary test showed no specific adsorption to the TOCNF−DHP MPs, with more than 90% and 30% of MB and Arg, respectively, adsorbed to TOCNF-free DHP MPs. TOCNF contains many carboxylate groups on the nanofiber surface [48], and the reported core–shell MPs containing 2.5% (w/v) TOCNF on the surface have been reported to show good drug loading capability via electrostatic interaction between TOCNF and ionic drugs [37]. In this study, the TOCNF–DHP MPs possess relatively low amounts of TOCNF, resulting in little difference in the drug-loading capacity. Nonetheless, cellulose and lignin are expected to exhibit drug loading/release capabilities, as previously reported [37,58,69], indicating that the further design of TOCNF–DHP MP materials might enhance drug loading by tuning the combination of TOCNF and DHP in the enzymatic green synthesis without using organic solvents. The long-term stability test [65] still needs to be done for practical availability. In this work, there was no big difference in the morphology of the MPs before and after the drug loading test. Such stability test in cosmetic applications will be investigated in a future work. Our enzymatic preparation of hybrid MPs from DHP and TOCNF can be applied for other combination, e.g., ferulic acid and naturally-occurring phenols [71,72] with various nanocellulose. At this stage, it may involve the difficulty in the large-scale production; however, such a green approach would be highly expected further in the future to design various nanomaterials from biomass-based natural biomolecules.

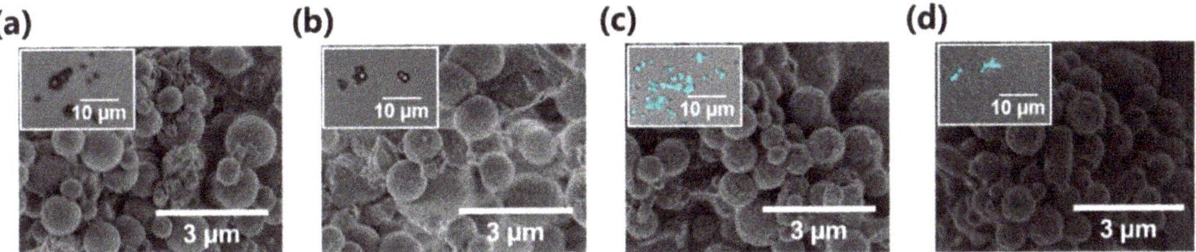

Figure 7. SEM and CLSM images (inset) of (**a,b**) DHP MPs and (**c,d**) TOCNF–DHP MPs (TOCNF, 0.5% (w/v)) after loading with (**a,c**) methylene blue (MB) and (**b,d**) arginine (Arg). CLSM images were obtained after Calcofluor white staining. Each inset is a merged picture of bright field and fluorescent images obtained by CLSM.

4. Conclusions

The successful green synthesis of DHP MPs functionalized with TOCNF was achieved in a one-pot one-step aqueous process without any organic solvents. Horseradish peroxidase enzymatically catalyzed the radical coupling of coniferyl alcohol in an aqueous suspension of TOCNFs, and the resultant TOCNF–DHP composites possessed a spherical shape and submicron-order size of ca. 0.8 µm in diameter, which would be expected to have cosmetic applications. The TOCNF–DHP MP surfaces were covered with TOCNF that possessed ζ-potential of ca. −40 mV, possibly improving the stability of particle dispersion. Some TOCNF was also embedded inside the MPs. As both TOCNF and lignin are polymers known to be biodegradable in the sea, TOCNF–DHP MPs have great potential as alternatives to nonbiodegradable MPs.

Author Contributions: N.F. and M.H. conducted all experiments and analytical characterization. T.K. conceived the presented idea and designed research. N.F. and T.K. contributed to writing the manuscript. All authors have read and approved the published version of the manuscript.

Funding: This research was funded by the Grant-in-Aid for Scientific Research (KAKENHI) Program (grant numbers JP20K22592 to M.H. and JP18K19233 to T.K.) from the Japan Society for the Promotion of Science, the Kyushu University-Initiated Venture Business Seed Development Program (GAP Fund to T.K.), and the Short-term Intensive Research Support Program from the Faculty of Agriculture, Kyushu University (M.H. and T.K.).

Institutional Review Board Statement: Not applicable.

Informed Consent Statement: Not applicable.

Data Availability Statement: Data presented in this study are available in this article.

Acknowledgments: The authors are very grateful to Midori Watanabe at the Center of Advanced Instrumental Analysis, Kyushu University, for providing technical support in SPM and SEM analyses, Yumi Fukunaga at the Ultramicroscopy Research Center, Kyushu University, for providing support in SEM sample preparation, and Yuki Okugawa at the Center for Advanced Instrumental and Educational Supports, Faculty of Agriculture, Kyushu University, for providing support in CLSM observation. We thank Simon Partridge, from Edanz Group for editing a draft of this manuscript.

Conflicts of Interest: The authors declare no conflict of interest.

References

1. Cole, M.; Lindeque, P.; Halsband, C.; Galloway, T.S. Microplastics as contaminants in the marine environment: A review. *Mar. Pollut. Bull.* **2011**, *62*, 2588–2597. [CrossRef]
2. Yee, M.S.L.; Hii, L.W.; Looi, C.K.; Lim, W.M.; Wong, S.F.; Kok, Y.Y.; Tan, B.K.; Wong, C.Y.; Leong, C.O. Impact of microplastics and nanoplastics on human health. *Nanomaterials* **2021**, *11*, 496. [CrossRef] [PubMed]
3. Browne, M.A.; Crump, P.; Niven, S.J.; Teuten, E.; Tonkin, A.; Galloway, T.; Thompson, R. Accumulation of microplastic on shorelines woldwide: Sources and sinks. *Environ. Sci. Technol.* **2011**, *45*, 9175–9179. [CrossRef] [PubMed]
4. Cheung, P.K.; Fok, L. Characterisation of plastic microbeads in facial scrubs and their estimated emissions in Mainland China. *Water Res.* **2017**, *122*, 53–61. [CrossRef] [PubMed]
5. Auta, H.S.; Emenike, C.U.; Fauziah, S.H. Distribution and importance of microplastics in the marine environment: A review of the sources, fate, effects, and potential solutions. *Environ. Int.* **2017**, *102*, 165–176. [CrossRef]
6. El-Habashy, S.E.; Eltaher, H.M.; Gaballah, A.; Zaki, E.I.; Mehanna, R.A.; El-Kamel, A.H. Hybrid bioactive hydroxyapatite/polycaprolactone nanoparticles for enhanced osteogenesis. *Mater. Sci. Eng. C* **2021**, *119*, 111599. [CrossRef]
7. Fan, X.; Zou, Y.; Geng, N.; Liu, J.; Hou, J.; Li, D.; Yang, C.; Li, Y. Investigation on the adsorption and desorption behaviors of antibiotics by degradable MPs with or without UV ageing process. *J. Hazard. Mater.* **2021**, *401*, 123363. [CrossRef]
8. Im, J.; Jang, E.K.; Yim, D.B.; Kim, J.H.; Cho, K.Y. One-pot fabrication of uniform half-moon-shaped biodegradable microparticles via microfluidic approach. *J. Ind. Eng. Chem.* **2020**, *90*, 152–158. [CrossRef]
9. Jiang, P.; Jacobs, K.M.; Ohr, M.P.; Swindle-Reilly, K.E. Chitosan-Polycaprolactone Core-Shell Microparticles for Sustained Delivery of Bevacizumab. *Mol. Pharm.* **2020**, *17*, 2570–2584. [CrossRef]
10. Ge, Y.; Dababneh, F.; Li, L. Economic Evaluation of Lignocellulosic Biofuel Manufacturing Considering Integrated Lignin Waste Conversion to Hydrocarbon Fuels. *Procedia Manuf.* **2017**, *10*, 112–122. [CrossRef]
11. Wang, Q.; Tian, D.; Hu, J.; Shen, F.; Yang, G.; Zhang, Y.; Deng, S.; Zhang, J.; Zeng, Y.; Hu, Y. Fates of hemicellulose, lignin and cellulose in concentrated phosphoric acid with hydrogen peroxide (PHP) pretreatment. *RSC Adv.* **2018**, *8*, 12714–12723. [CrossRef]
12. Funahashi, R.; Okita, Y.; Hondo, H.; Zhao, M.; Saito, T.; Isogai, A. Different Conformations of Surface Cellulose Molecules in Native Cellulose Microfibrils Revealed by Layer-by-Layer Peeling. *Biomacromolecules* **2017**, *18*, 3687–3694. [CrossRef] [PubMed]
13. Daicho, K.; Saito, T.; Fujisawa, S.; Isogai, A. The Crystallinity of Nanocellulose: Dispersion-Induced Disordering of the Grain Boundary in Biologically Structured Cellulose. *ACS Appl. Nano Mater.* **2018**, *1*, 5774–5785. [CrossRef]
14. Moon, R.J.; Martini, A.; Nairn, J.; Simonsen, J.; Youngblood, J. Cellulose nanomaterials review: Structure, properties and nanocomposites. *Chem. Soc. Rev.* **2011**, *40*, 3941–3994. [CrossRef] [PubMed]
15. Quinlan, R.J.; Sweeney, M.D.; Lo Leggio, L.; Otten, H.; Poulsen, J.C.N.; Johansen, K.S.; Krogh, K.B.R.M.; Jørgensen, C.I.; Tovborg, M.; Anthonsen, A.; et al. Insights into the oxidative degradation of cellulose by a copper metalloenzyme that exploits biomass components. *Proc. Natl. Acad. Sci. USA* **2011**, *108*, 15079–15084. [CrossRef]
16. Tamo, A.K.; Doench, I.; Helguera, A.M.; Hoenders, D.; Walther, A.; Madrazo, A.O. Biodegradation of crystalline cellulose nanofibers by means of enzyme immobilized-alginate beads and microparticles. *Polymers* **2020**, *12*, 1522. [CrossRef]
17. Shen, Y.; Li, Z.; Huo, Y.Y.; Bao, L.; Gao, B.; Xiao, P.; Hu, X.; Xu, X.W.; Li, J. Structural and Functional Insights Into CmGH1, a Novel GH39 Family β-Glucosidase From Deep-Sea Bacterium. *Front. Microbiol.* **2019**, *10*, 2922. [CrossRef]
18. Achyuthan, K.E.; Achyuthan, A.M.; Adams, P.D.; Dirk, S.M.; Harper, J.C.; Simmons, B.A.; Singh, A.K. Supramolecular self-assembled chaos: Polyphenolic lignin's barrier to cost-effective lignocellulosic biofuels. *Molecules* **2010**, *15*, 8641–8688. [CrossRef]
19. Simon, C.; Spriet, C.; Hawkins, S.; Lion, C. Visualizing lignification dynamics in plants with click chemistry: Dual labeling is BLISS! *J. Vis. Exp.* **2018**, *131*, e56947. [CrossRef]
20. Janusz, G.; Pawlik, A.; Sulej, J.; Świderska-Burek, U.; Jarosz-Wilkolazka, A.; Paszczyński, A. Lignin degradation: Microorganisms, enzymes involved, genomes analysis and evolution. *FEMS Microbiol. Rev.* **2017**, *41*, 941–962. [CrossRef] [PubMed]
21. Datta, R.; Kelkar, A.; Baraniya, D.; Molaei, A.; Moulick, A.; Meena, R.S.; Formanek, P. Enzymatic degradation of lignin in soil: A review. *Sustainability* **2017**, *9*, 1163. [CrossRef]
22. Ohta, Y.; Hasegawa, R.; Kurosawa, K.; Maeda, A.H.; Koizumi, T.; Nishimura, H.; Okada, H.; Qu, C.; Saito, K.; Watanabe, T.; et al. Enzymatic Specific Production and Chemical Functionalization of Phenylpropanone Platform Monomers from Lignin. *ChemSusChem* **2017**, *10*, 425–433. [CrossRef] [PubMed]
23. Thomas, B.; Raj, M.C.; Athira, B.K.; Rubiyah, H.M.; Joy, J.; Moores, A.; Drisko, G.L.; Sanchez, C. Nanocellulose, a Versatile Green Platform: From Biosources to Materials and Their Applications. *Chem. Rev.* **2018**, *118*, 11575–11625. [CrossRef] [PubMed]

24. Yang, X.; Biswas, S.K.; Han, J.; Tanpichai, S.; Li, M.C.; Chen, C.; Zhu, S.; Das, A.K.; Yano, H. Surface and interface engineering for nanocellulosic advanced materials. *Adv. Mater.* **2020**, *2002264*. [CrossRef]

25. Figueiredo, P.; Lintinen, K.; Hirvonen, J.T.; Kostiainen, M.A.; Santos, H.A. Properties and chemical modifications of lignin: Towards lignin-based nanomaterials for biomedical applications. *Prog. Mater. Sci.* **2018**, *93*, 233–269. [CrossRef]

26. Lu, Y.; Han, J.; Ding, Q.; Yue, Y.; Xia, C.; Ge, S.; Van Le, Q.; Dou, X.; Sonne, C.; Lam, S.S. TEMPO-oxidized cellulose nanofibers/polyacrylamide hybrid hydrogel with intrinsic self-recovery and shape memory properties. *Cellulose* **2021**, *8*, 1469–1488. [CrossRef]

27. Iwatake, A.; Nogi, M.; Yano, H. Cellulose nanofiber-reinforced polylactic acid. *Compos. Sci. Technol.* **2008**, *68*, 2103–2106. [CrossRef]

28. Niu, X.; Liu, Y.; Song, Y.; Han, J.; Pan, H. Rosin modified cellulose nanofiber as a reinforcing and co-antimicrobial agents in polylactic acid/chitosan composite film for food packaging. *Carbohydr. Polym.* **2018**, *183*, 102–109. [CrossRef]

29. Kanomata, K.; Fukuda, N.; Miyata, T.; Lam, P.Y.; Takano, T.; Tobimatsu, Y.; Kitaoka, T. Lignin-Inspired Surface Modification of Nanocellulose by Enzyme-Catalyzed Radical Coupling of Coniferyl Alcohol in Pickering Emulsion. *ACS Sustain. Chem. Eng.* **2020**, *8*, 1185–1194. [CrossRef]

30. Gao, H.; Duan, B.; Lu, A.; Deng, H.; Du, Y.; Shi, X.; Zhang, L. Fabrication of cellulose nanofibers from waste brown algae and their potential application as milk thickeners. *Food Hydrocoll.* **2018**, *79*, 473–481. [CrossRef]

31. Ullah, H.; Santos, H.A.; Khan, T. Applications of bacterial cellulose in food, cosmetics and drug delivery. *Cellulose* **2016**, *23*, 2291–2314. [CrossRef]

32. Yadav, H.M.; Park, J.D.; Kang, H.C.; Kim, J.; Lee, J.J. Cellulose nanofiber composite with bimetallic zeolite imidazole framework for electrochemical supercapacitors. *Nanomaterials* **2021**, *11*, 395. [CrossRef] [PubMed]

33. Gopakumar, D.A.; Pai, A.R.; Pottathara, Y.B.; Pasquini, D.; Carlos De Morais, L.; Luke, M.; Kalarikkal, N.; Grohens, Y.; Thomas, S. Cellulose nanofiber-based polyaniline flexible papers as sustainable microwave absorbers in the X-band. *ACS Appl. Mater. Interfaces* **2018**, *10*, 20032–20043. [CrossRef] [PubMed]

34. Yan, C.; Wang, J.; Kang, W.; Cui, M.; Wang, X.; Foo, C.Y.; Chee, K.J.; Lee, P.S. Highly stretchable piezoresistive graphene-nanocellulose nanopaper for strain sensors. *Adv. Mater.* **2014**, *26*, 2022–2027. [CrossRef] [PubMed]

35. Yeasmin, S.; Yeum, J.H.; Ji, B.C.; Choi, J.H. Electrically Conducting Pullulan-Based Nanobiocomposites Using Carbon Nanotubes and TEMPO Cellulose Nanofibril. *Nanomaterials* **2021**, *11*, 602. [CrossRef]

36. Fujisawa, S.; Kaku, Y.; Kimura, S.; Saito, T. Magnetically Collectable Nanocellulose-Coated Polymer Microparticles by Emulsion Templating. *Langmuir* **2020**, *36*, 9235–9240. [CrossRef]

37. Fujisawa, S.; Togawa, E.; Kuroda, K.; Saito, T.; Isogai, A. Fabrication of ultrathin nanocellulose shells on tough microparticles via an emulsion-templated colloidal assembly: Towards versatile carrier materials. *Nanoscale* **2019**, *11*, 15004–15009. [CrossRef]

38. Zhang, B.; Zhang, Z.; Kapar, S.; Ataeian, P.; Da Silva Bernardes, J.; Berry, R.; Zhao, W.; Zhou, G.; Tam, K.C. Microencapsulation of Phase Change Materials with Polystyrene/Cellulose Nanocrystal Hybrid Shell via Pickering Emulsion Polymerization. *ACS Sustain. Chem. Eng.* **2019**, *7*, 17756–17767. [CrossRef]

39. Fujisawa, S.; Togawa, E.; Kuroda, K. Facile Route to Transparent, Strong, and Thermally Stable Nanocellulose/Polymer Nanocomposites from an Aqueous Pickering Emulsion. *Biomacromolecules* **2017**, *18*, 266–271. [CrossRef] [PubMed]

40. Beisl, S.; Miltner, A.; Friedl, A. Lignin from micro- to nanosize: Production methods. *Int. J. Mol. Sci.* **2017**, *18*, 2367. [CrossRef] [PubMed]

41. Mishra, P.K.; Ekielski, A. The self-assembly of lignin and its application in nanoparticle synthesis: A short review. *Nanomaterials* **2019**, *9*, 243. [CrossRef]

42. Cathala, B.; Saake, B.; Faix, O.; Monties, B. Evaluation of the reproducibility of the synthesis of dehydrogenation polymer models of lignin. *Polym. Degrad. Stab.* **1998**, *59*, 65–69. [CrossRef]

43. Touzel, J.P.; Chabbert, B.; Monties, B.; Debeire, P.; Cathala, B. Synthesis and characterization of dehydrogenation polymers in Gluconacetobacter xylinus cellulose and cellulose/pectin composite. *J. Agric. Food Chem.* **2003**, *51*, 981–986. [CrossRef]

44. Mićić, M.; Jeremić, M.; Radotić, K.; Leblanc, R.M. A comparative study of enzymatically and photochemically polymerized artificial lignin supramolecular structures using environmental scanning electron microscopy. *J. Colloid Interface Sci.* **2000**, *231*. [CrossRef]

45. Micic, M.; Radotic, K.; Benitez, I.; Ruano, M.; Jeremic, M.; Moy, V.; Mabrouki, M.; Leblanc, R.M. Topographical characterization and surface force spectroscopy of the photochemical lignin model compound. *Biophys. Chem.* **2001**, *94*, 257–263. [CrossRef]

46. Muraille, L.; Aguié-Béghin, V.; Chabbert, B.; Molinari, M. Bioinspired lignocellulosic films to understand the mechanical properties of lignified plant cell walls at nanoscale. *Sci. Rep.* **2017**, *7*, 44065. [CrossRef] [PubMed]

47. Barone, J.R. Composites of Nanocellulose and Lignin-like Polymers. *Cellul. Based Compos. New Green Nanomater.* **2014**, *9783527327*, 185–200.

48. Saito, T.; Nishiyama, Y.; Putaux, J.L.; Vignon, M.; Isogai, A. Homogeneous suspensions of individualized microfibrils from TEMPO-catalyzed oxidation of native cellulose. *Biomacromolecules* **2006**, *7*, 1687–1691. [CrossRef] [PubMed]

49. Jiang, J.; Chen, H.; Liu, L.; Yu, J.; Fan, Y.; Saito, T.; Isogai, A. Influence of chemical and enzymatic TEMPO-mediated oxidation on chemical structure and nanofibrillation of lignocellulose. *ACS Sustain. Chem. Eng.* **2020**, *8*, 14198–14206. [CrossRef]

50. Kisszekelyi, P.; Hardian, R.; Vovusha, H.; Chen, B.; Zeng, X.; Schwingenschlögl, U.; Kupai, J.; Szekely, G. Selective electrocatalytic oxidation of biomass-derived 5-hydroxymethylfurfural to 2,5-diformylfuran: From mechanistic investigations to catalyst recovery. *ChemSusChem* **2020**, *13*, 3127–3136. [CrossRef] [PubMed]

51. Akhlaghi, Y.; Ghaffari, S.; Attar, H.; Alamir Hoor, A. A rapid hydrolysis method and DABS-Cl derivatization for complete amino acid analysis of octreotide acetate by reversed phase HPLC. *Amino Acids* **2015**, *47*, 2255–2263. [CrossRef] [PubMed]

52. Sipponen, M.H.; Lange, H.; Ago, M.; Crestini, C. Understanding Lignin Aggregation Processes. A Case Study: Budesonide Entrapment and Stimuli Controlled Release from Lignin Nanoparticles. *ACS Sustain. Chem. Eng.* **2018**, *6*, 9342–9351. [CrossRef] [PubMed]

53. Moberg, T.; Sahlin, K.; Yao, K.; Geng, S.; Westman, G.; Zhou, Q.; Oksman, K.; Rigdahl, M. Rheological properties of nanocellulose suspensions: Effects of fibril/particle dimensions and surface characteristics. *Cellulose* **2017**, *24*, 2499–2510. [CrossRef]

54. Bock, P.; Nousiainen, P.; Elder, T.; Blaukopf, M.; Amer, H.; Zirbs, R.; Potthast, A.; Gierlinger, N. Infrared and Raman spectra of lignin substructures: Dibenzodioxocin. *J. Raman Spectrosc.* **2020**, *51*, 422–431. [CrossRef]

55. Lin, C.C.; Lin, W.J. Sun protection factor analysis of sunscreens containing titanium dioxide nanoparticles. *J. Food Drug Anal.* **2011**, *19*, 1–8. [CrossRef]

56. Goi, Y.; Fujisawa, S.; Saito, T.; Yamane, K.; Kuroda, K.; Isogai, A. Dual Functions of TEMPO-Oxidized Cellulose Nanofibers in Oil-in-Water Emulsions: A Pickering Emulsifier and a Unique Dispersion Stabilizer. *Langmuir* **2019**, *35*, 10920–10926. [CrossRef] [PubMed]

57. Mussatto, A.; Groarke, R.; O'Neill, A.; Obeidi, M.A.; Delaure, Y.; Brabazon, D. Influences of powder morphology and spreading parameters on the powder bed topography uniformity in powder bed fusion metal additive manufacturing. *Addit. Manuf.* **2021**, *38*, 101807.

58. Dai, L.; Liu, R.; Hu, L.Q.; Zou, Z.F.; Si, C.L. Lignin Nanoparticle as a Novel Green Carrier for the Efficient Delivery of Resveratrol. *ACS Sustain. Chem. Eng.* **2017**, *5*, 8241–8249. [CrossRef]

59. Bhattacharjee, S. DLS and zeta potential-What they are and what they are not? *J. Control. Release* **2016**, *235*, 337–351. [CrossRef]

60. Park, J.Y.; Park, C.W.; Han, S.Y.; Kwon, G.J.; Kim, N.H.; Lee, S.H. Effects of pH on nanofibrillation of TEMPO-oxidized paper mulberry bast fibers. *Polymers* **2019**, *11*, 414. [CrossRef]

61. Wu, Y.; Qian, Y.; Zhang, A.; Lou, H.; Yang, D.; Qiu, X. Light Color Dihydroxybenzophenone Grafted Lignin with High UVA/UVB Absorbance Ratio for Efficient and Safe Natural Sunscreen. *Ind. Eng. Chem. Res.* **2020**, *59*, 17057–17068. [CrossRef]

62. Micic, M.; Radotic, K.; Jeremic, M.; Djikanovic, D.; Kämmer, S.B. Study of the lignin model compound supramolecular structure by combination of near-field scanning optical microscopy and atomic force microscopy. *Colloids Surf. B Biointerfaces* **2004**, *34*, 33–40. [CrossRef] [PubMed]

63. Da Silva Ferez, D.; Ruggiero, R.; Morais, L.C.; Machado, A.E.H.; Mazea, K. Theoretical and experimental studies on the adsorption of aromatic compounds onto cellulose. *Langmuir* **2004**, *20*, 3151–3158. [CrossRef] [PubMed]

64. Tarasov, D.; Leitch, M.; Fatehi, P. Lignin-carbohydrate complexes: Properties, applications, analyses, and methods of extraction: A review. *Biotechnol. Biofuels* **2018**, *11*, 269. [CrossRef] [PubMed]

65. Anchisi, C.; Meloni, M.C.; Maccioni, A.M. Chitosan beads loaded with essential oils in cosmetic formulations. *J. Cosmet. Sci.* **2006**, *57*, 205–214. [CrossRef]

66. Manca, M.L.; Castangia, I.; Zaru, M.; Nácher, A.; Valenti, D.; Fernàndez-Busquets, X.; Fadda, A.M.; Manconi, M. Development of curcumin loaded sodium hyaluronate immobilized vesicles (hyalurosomes) and their potential on skin inflammation and wound restoring. *Biomaterials* **2015**, *71*, 100–109. [CrossRef]

67. Zhou, Y.; Qian, Y.; Wang, J.; Qiu, X.; Zeng, H. Bioinspired lignin-polydopamine nanocapsules with strong bioadhesion for long-acting and high-performance natural sunscreens. *Biomacromolecules* **2020**, *21*, 3231–3241. [CrossRef] [PubMed]

68. Fertah, M.; Belfkira, A.; Taourirte, M.; Brouillette, F. Controlled release of diclofenac by a new system based on a cellulosic substrate and calcium alginate. *BioResources* **2015**, *10*, 5932–5948. [CrossRef]

69. Chin, S.F.; Jimmy, F.B.; Pang, S.C. Size controlled fabrication of cellulose nanoparticles for drug delivery applications. *J. Drug Deliv. Sci. Technol.* **2018**, *43*, 262–266. [CrossRef]

70. Maeno, K. Direct Quantification of Natural Moisturizing Factors in Stratum Corneum using Direct Analysis in Real Time Mass Spectrometry with Inkjet-Printing Technique. *Sci. Rep.* **2019**, *9*, 17789. [CrossRef]

71. Izawa, H.; Miyazaki, Y.; Ifuku, S.; Morimoto, M.; Saimoto, H. Fully biobased oligophenolic nanoparticle prepared by horseradish peroxidase-catalyzed polymerization. *Chem. Lett.* **2016**, *45*, 631–633. [CrossRef]

72. Li, Z.; Renneckar, S.; Barone, J.R. Nanocomposites prepared by in situ enzymatic polymerization of phenol with TEMPO-oxidized nanocellulose. *Cellulose* **2010**, *17*, 57–68. [CrossRef]

 *nanomaterials*

Article

Highly Stable Pickering Emulsions with Xylan Hydrate Nanocrystals

Shanyong Wang and Zhouyang Xiang *

State Key Laboratory of Pulp and Paper Engineering, South China University of Technology, Guangzhou 510640, China; feshanyong@mail.scut.edu.cn
* Correspondence: fezyxiang@scut.edu.cn; Tel.: +86-20-87113753

Abstract: Xylan is a highly abundant plant-based biopolymer. Original xylans in plants are in an amorphous state, but deacetylated and low-branched xylan can form a crystalline structure with water molecules. The utilizations of xylan have been limited to bulk applications either with inconsistency and uncertainty or with extensive chemical derivatization due to the insufficient studies on its crystallization. The applications of xylan could be greatly broadened in advanced green materials if xylan crystals are effectively utilized. In this paper, we show a completely green production of nano-sized xylan crystals and propose their application in forming Pickering emulsions. The branches of xylan were regulated during the separation step to controllably induce the formation of xylan hydrate crystals. Xylan hydrate nanocrystals (XNCs) with a uniform size were successfully produced solely by a mild ultrasonic treatment. XNCs can be adsorbed onto oil–water interfaces at a high density to form highly stable Pickering emulsions. The emulsifying properties of XNCs were comparable to some synthetic emulsifiers and better than some other common biopolymer nanocrystals, demonstrating that XNCs have great potential in industrial emulsification.

Keywords: hemicellulose; xylan; nanocrystal; Pickering emulsion

Citation: Wang, S.; Xiang, Z. Highly Stable Pickering Emulsions with Xylan Hydrate Nanocrystals. *Nanomaterials* **2021**, *11*, 2558. https://doi.org/10.3390/nano11102558

Academic Editor: Takuya Kitaoka

Received: 30 August 2021
Accepted: 26 September 2021
Published: 29 September 2021

Publisher's Note: MDPI stays neutral with regard to jurisdictional claims in published maps and institutional affiliations.

1. Introduction

Hemicelluloses are a type of very abundant plant-based biopolymers [1,2]. Xylan is a major type of hemicellulose and is primarily found in hardwoods and graminaceous plants. Xylan is also present in softwoods but in relatively low amounts. Depending on the plant species, xylans have β-(1-4)-linked D-xylopyranosyl backbones but are substituted with one or more monomers of L-arabinofuranosyl, D-glucuronopyranosyl, (4-O-methyl)-D-glucuronopyranosyl, and some minor residues [3]. The backbones of xylans in hardwood and graminaceous plants are partially acetylated, but softwood xylans are not acetylated. Original xylans in plants are in an amorphous state [4], but deacetylated and low-branched xylans can form a crystalline structure with solvent molecules, such as water and dimethyl sulfoxide (DMSO) [5–7]. Due to the insufficient studies on xylan crystallization, the applications of xylans have been mainly focused on bulk applications, such as films and coatings [8,9]. However, xylan crystallization affects its solubility and viscosity [10] and is detrimental to its water cast film formation [6], adding inconsistency and uncertainty to its bulk applications. The applications of xylan could be greatly broadened if xylan crystals are nano-sized, as nanocrystals of some biopolymers, e.g., cellulose, chitin/chitosan, and starch, have been widely used in advanced functional materials, such as biological scaffolds, drug carriers, optical materials, permselective membranes, polymer electrolytes, and emulsion stabilizers [11,12]. However, preparing xylan nano-crystals with controllable size and shape is difficult and has only been reported on in a few studies [7,13]. Chanzy et al. reported the desizes of xylan crystals by enzymatic degradation, but the enzyme degrades the xylan crystals from the edge and destroys the crystalized structure [13]. Wang et al. reported the nano-sizing of xylan hydrate crystals by surface carboxymethylation, but this method resulted in a significant decrease in crystallinity and uncontrollable sizes [14]. Meng et al.

recently found that well-dispersed alkali-extracted xylan in DMSO can spontaneously form xylan–DMSO crystal nanowhiskers with a controllable size by heating induction [7]; the mechanisms behind it were still not fully understood and the application of the xylan–DMSO crystal in an aqueous environment is not known. The limited knowledge of the controllable preparation of nano-sized xylan crystals has largely impeded its applications in advanced materials.

Many of the bio-based nanocrystals, e.g., cellulose nanocrystals (CNC) [15] and chitin nanocrystals [16], require severe acid treatments or chemical derivatizations to produce, causing chemical recycling and pollution issues. Due to a much less bulky structure of xylan, compared to cellulose and chitin, we hypothesized that the nano-sizing of xylan hydrate crystals to produce xylan hydrate nanocrystals (XNCs) only requires a mild physical treatment with a low energy input, which would be a completely green process. However, xylan is not always in a crystalized form [6], so controllably separating xylans with a low degree of branching to induce crystallization is an important prerequisite to the production of XNCs.

CNCs [15,17,18], chitin nanocrystals [16,19], and starch nanocrystals [20,21] have already been reported to be effective in forming and stabilizing Pickering emulsions. During the emulsifying process, the nanocrystals were adsorbed onto the oil–water interface, irreversibly, and formed a highly rigid layer to prevent the aggregation between discontinuous phases [12]. However, the application of XNCs in the Pickering emulsion has never been reported. Since the crystallization of xylans requires the involvement of solvent molecules, they have a high stability in the corresponding solvents [5,7,22]. Xylan hydrate crystals are very stable in water, as demonstrated by their water insolubility [6,10], so their nano-sized form is hypothesized to be especially suitable in the forming of oil–water Pickering emulsions.

2. Experimental Section

2.1. Material

Glacial acetic acid (AR), sodium chlorite (80%), sodium hydroxide (AR), sodium dodecyl sulfate (SDS, AR), Tween 80 (T80, pharmaceutical grade), Triton X-100 (TX100, AR), gum arabic (GA, pharmaceutical grade), fluorescent brightener VBL, styrene, and 2,2′-Azobis (2-methylpropionitrile) (ABIN, AR) were purchased from Shanghai Macklin Biochemical Co., Ltd. (Shanghai, China) Medium-chain triglycerides (the carbon number of each fatty acid chain is 8–10) were purchased from Shanghai Yuanye Bio-Technology Co., Ltd. (Shanghai, China) Absolute ethanol (AR) and hydrochloric acid were of analytical grade.

2.2. Separation and Characterization of Xylan

A grounded sugarcane bagasse sample, sieved between 40 and 60 mesh, was extracted in a Soxhlet extractor with 95% ethanol for 6 h to remove extractives, fat, and wax [6]. Holocelluloses were separated by treating the bagasse in a 2 wt% sodium chlorite solution with pH = 3.9–4 and a solid–liquid ratio of 1:20 at 75 °C for 1 h. The holocelluloses were washed three times with deionized water and one time with absolute ethanol and then air dried.

Xylans were separated from holocelluloses by alkaline extraction or successive alkaline extraction with NaOH solutions of different concentrations at 90 °C for 3 h (Table 1) [6,23]. After the extraction, the filtrate was adjusted to pH 5.5–6 by HCl and then concentrated to one-third of the original volume. The filtrate was then added to absolute ethanol with volumes three times of the filtrate and then stood overnight. The precipitate was centrifuged (Centrisart G-16, Sartorius, Gottingen, Germany) at 4200 rpm for 10 min, washed with 70% ethanol (*v:v*) three times, freeze dried, and ground to powders. The white powders obtained were xylans.

Table 1. Xylans separated through different extraction methods.

Xylan Fractions	Extraction Methods	Yield %
X1	1 wt% NaOH	10.77 ± 0.21
X2	2 wt% NaOH	21.55 ± 0.10
X3	4 wt% NaOH (successive after 2 wt%)	7.71 ± 0.04
X4	6 wt% NaOH (successive after 2 and 4 wt%)	9.08 ± 0.12

Note: the yields of xylans corresponded only to the indicated extraction step for successive extraction and were calculated based on the starting holocellulose.

The xylans were hydrolyzed according to the protocols described previously [24] in order to determine their lignin, neutral sugar, and uronic acid compositions. Neutral sugars were quantified by an ion chromatography system (IC-3000, Dionex, Sunnyvale, CA, USA) equipped with an anion-exchange column (CarboPacTM PA20, Dionex, Sunnyvale, CA, USA); the column was pre-equilibrated with 200 mM NaOH and gradiently eluted with water, 20 mM NaOH, and 500 mM sodium acetate at a flow rate of 0.5 mL/min. The estimation of the uronic acid content was based on the photometric method developed by Filisetti-Cozzi [25]. The molecular weight distribution of xylans was measured by GPC (PL-GPC50, Agilent Technologies, Santa Clara, CA, USA), with DMSO as the mobile phase.

The morphology of xylan particles in suspension was observed with a polarized microscope (BX53M, Olympus, Tokyo, Japan) equipped with a U-AN360P analyzer slider. X-ray diffraction (XRD) measurements on xylan powders were performed in the symmetric reflection mode using X'Pert Powder (Panalytical, Eindhoven, The Netherlands) equipped with a copper X-ray source (k = 1.5418 Å) and operated at 40 kV and 40 mA.

2.3. Preparation and Characterization of XNCs

The xylan aqueous suspensions were sonicated at 15 W/mL by an ultrasonic cell grinder (JY99-IIDN, Ningbo Xinzhi Biological Co., Ltd., Ningbo, China) for 40 min to obtain a stable XNC colloidal dispersion. The XNC colloidal dispersion was diluted 500 times and then dropped and air dried on a mica sheet, which was analyzed by an atomic force microscope (AFM, Nanoscope IIIa, Bruck, Germany). The particle size distribution of XNCs was calculated from the AFM images using Nanoscope Analysis 1.7. The diluted XNC colloidal dispersion was stained with phosphotungstic acid and then analyzed by transmission electron microscopy (TEM). The viscosity of the XNC dispersion was measured by a digital rotational viscometer (LV-SSR, Shanghai Fangrui Instrument Co., Ltd., Shanghai, China).

2.4. Preparation and Characterization of Pickering Emulsion

Emulsions were prepared with triglycerides at an oil–water ratio of 2:8 under different emulsifier concentrations (0.5, 1, 2 and 4 wt%) and was homogenized by an ultrasonic cell grinder at 10 W/mL for 90 s. The emulsifying activity (EA), emulsion cream index (ECI), and emulsion droplet size were evaluated. The EA was measured according to the method of Hu [26]. Approximately 30 μL of emulsion was mixed with 15 mL in a 0.1 wt% SDS aqueous solution, and the absorbance was measured at 500 nm by a UV–VIS spectrophotometer (Shanghai Mapada Instrument Co., Ltd., Shanghai, China). The ECI was calculated based on the previous methods [17,18,26] with slight modifications. The prepared emulsion was continuously centrifuged at 4000 rpm, the layering of the emulsion at different centrifugation times was recorded, and the *ECI* was calculated as follows:

$$ECI = \frac{H_A}{H_T} \times 100\%$$

where H_A is the height of the upper phase, and H_T is the total height of emulsion. The emulsion droplet size was measured by a laser particle size analyzer (LA960S, HORIBA, Kyoto, Japan) at a refractive index of 1.5 and a solid content of 0.5 wt%.

The aqueous solution of fluorescent brightener (1 mg/mL) was used to dye the XNC colloidal dispersions (the volume ratio of brightener solution to dispersion was 1:100). The dyed XNC colloidal dispersion was used to prepare the Pickering emulsions. The fluorescence distribution of XNCs in the emulsion was observed by a laser confocal microscope (TCS SP5, Leica, Munich, Germany) at an excitation wavelength of 403 nm and a collection wavelength of 420–500 nm. According to previous conditions [17], the oil phase of the emulsion was replaced by styrene that was initiated with polymerization. A field-emission scanning electron microscope (FESEM, Merlin, Zeiss, Oberkohen, Germany) was used to observe the adsorption of XNCs on the surface of polystyrene droplets. In order to observe the emulsion droplet surface more easily, the dyed emulsions and the polymerized emulsions were dispersed by a lab high-speed homogenizer (T18, IKA, Guangzhou, China) instead of using ultrasonic treatment to obtain larger oil droplets.

3. Results and Discussion

3.1. Xylan Crystallinity

The chemical structures of xylans were regulated by alkaline extraction or successive alkaline extraction with NaOH solutions of different concentrations, giving xylan samples X1–X4 (Table 1). A high concentration of NaOH solution extracted xylans with less branches and a relatively lower molecular weight (X3 and X4), compared to those extracted by a low concentration of NaOH solution (Tables 2 and 3; Figure 1a), which was consistent with previous studies [6].

Table 2. Chemical compositions of the xylan separated through different extraction methods.

Samples	Arabinose %	Galactose %	Glucose %	Xylose %	Uronic Acid %	Lignin%
X1	12.72 ± 0.06	2.04 ± 0.07	5.59 ± 0.05	51.29 ± 0.36	6.77 ± 0.14	7.60 ± 1.55
X2	10.99 ± 0.38	2.18 ± 0.00	7.06 ± 0.12	67.12 ± 1.01	2.77 ± 0.08	1.54 ± 0.11
X3	6.66 ± 0.03	0.10 ± 0.08	6.06 ± 0.01	81.42 ± 0.64	2.24 ± 0.01	1.19 ± 0.51
X4	1.63 ± 0.16	0.27 ± 0.68	6.84 ± 2.79	74.56 ± 0.94	1.50 ± 0.01	0.88 ± 0.05

Table 3. Molecular structures of separated xylan fractions.

Samples	Mn	Mw	Polydispersity Index	* Degree of Substitution
X1	54,400	85,100	1.56	0.38
X2	52,800	95,400	1.81	0.21
X3	45,900	75,200	1.64	0.11
X4	31,300	51,400	1.64	0.04

* Calculated by the content ratios of (arabinose + uronic acid)/xylose.

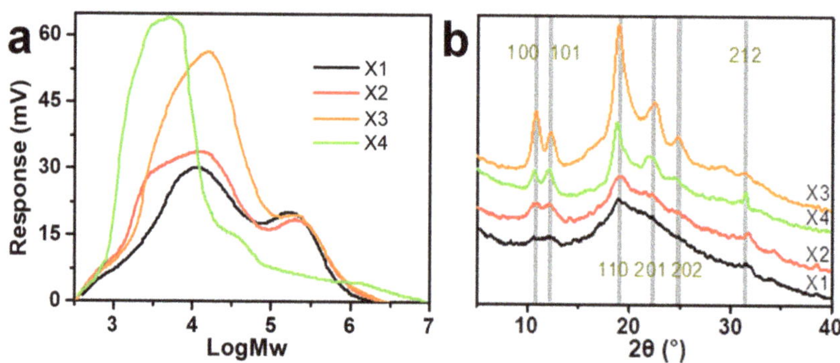

Figure 1. (**a**) Molecular weight distribution and (**b**) XRD profiles of different xylan samples.

The XRD profiles of the four xylan samples are shown in Figure 1b. X1 and X2 had profiles showing mostly amorphous states, while X3 and X4 had peaks characterizing typical xylan hydrate crystals [6]. The Miller indices of major peaks based on the trigonal unit cell of the xylan hydrate crystal in the study of Nieduszynski (1972) [5] are indicated. When observed under crossed polarizers, the suspension of X1 and X2 showed little birefringence, while X3 and X4 showed a large number of birefringent granular objects (Figure S1), further suggesting the high crystallinity of X3 and X4 samples. The crystallinity of xylan was positively correlated with the degree of branching, which is consistent with previous studies [6,27]. The crystallinity of X4 was slightly lower than that of X3, which may be related to the degradation of the X4 molecular chain (Table 2 and Figure 1a). Previous studies have shown that reducing the length of the polymer chains would increase their degree of freedom and, subsequently, reduce their possibility of aggregating into ordered structures [28], which indicates that the low molecular weight might not be conducive to the crystallization of xylan. These results indicated that regulating the degree of branching to induce the formation of xylan hydrate crystals was successful, facilitating the subsequent nano-sizing of xylan hydrate crystals.

3.2. XNC Preparation

Ultrasonic waves were introduced to the suspensions of xylan crystals in order to nano-size the xylan hydrate crystals to xylan hydrate nanocrystals (XNCs). The power of the ultrasonic treatment was carefully controlled at a mild level, because a higher power would have destroyed the crystals and a lower power would not have been able to nano-size the crystals. Before the ultrasonic treatment, xylans X1 and X2 were dissolved in water into a transparent solution upon heating, while X3 and X4 could not form stable suspensions but deposited at the bottom of water (Figure S2), again suggesting that xylan solubility relates to its crystallinity (Figure 1b). After ultrasonic treatment, the evident optical path of the red laser through the treated suspensions indicated the formation of stable colloidal dispersions UX1-4 (Figure 2a). AFM and TEM images of UX3 and UX4 demonstrated nanoparticles with high uniformity and a spindle shape, indicating that uniformly sized XNCs had been successfully prepared (Figure 2b,c). UX4 showed a particle diameter (~45.5 nm) larger than that of UX3 (~21.5 nm); UX4 also showed a wider particle size distribution than that of UX3 (Figure 3). UX1 and UX2 both showed a few shapeless particles in the field of vision, a very narrow particle size distribution, and a small particle size, suggesting the dissolution or degradation of noncrystalline xylan particles (Figures 2b and 3). The diameter of XNCs showed a positive correlation to the crystallinity of starting xylan samples (Figure 1b).

In sum, there are mostly noncrystalline and soluble xylans in UX1 and UX2 colloidal dispersions, while there are mostly XNC in UX3 and UX4 colloidal dispersions. The XNC is a bio-based nano-material similar to the cellulose nanocrystal (CNC). Different from cellulose, xylan is not always in a crystalized form [6], so controllably separating xylans with a low degree of branching to induce crystallization is an important prerequisite to the production of XNCs. However, the production of CNCs requires a concentrated acid treatment. Due to the less bulky structure of xylan, compared to cellulose, XNC production only requires ultrasonic treatment with a mild energy input, which is completely green without any acid treatments and chemical derivatizations.

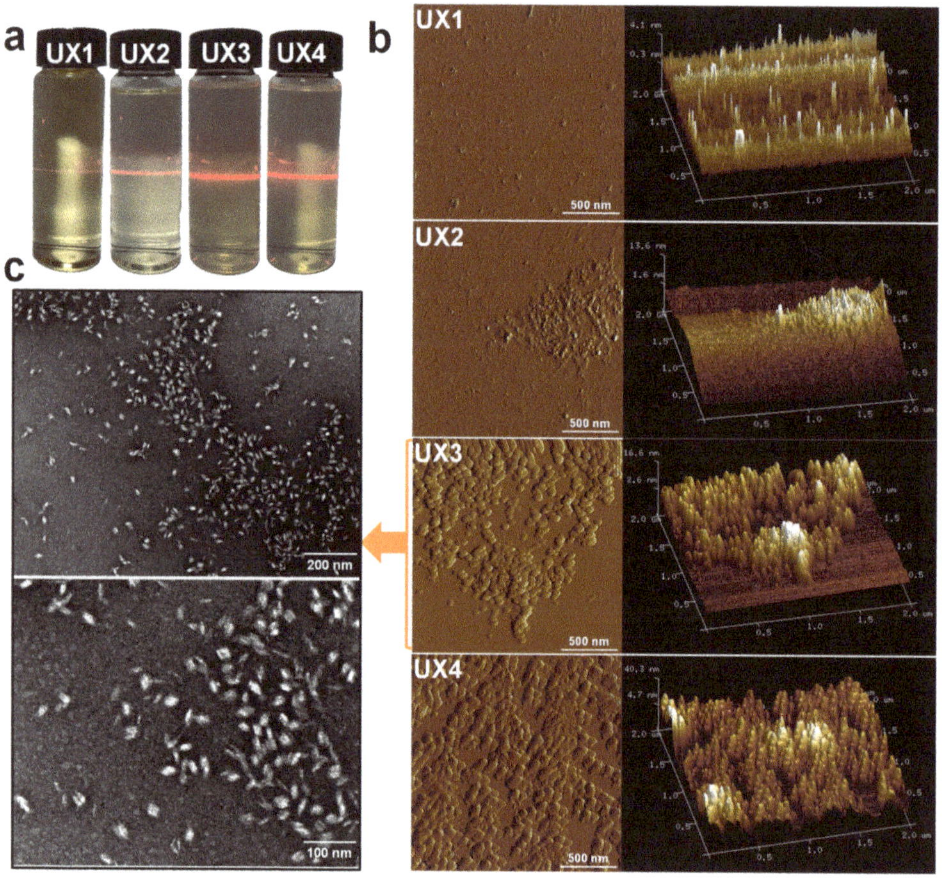

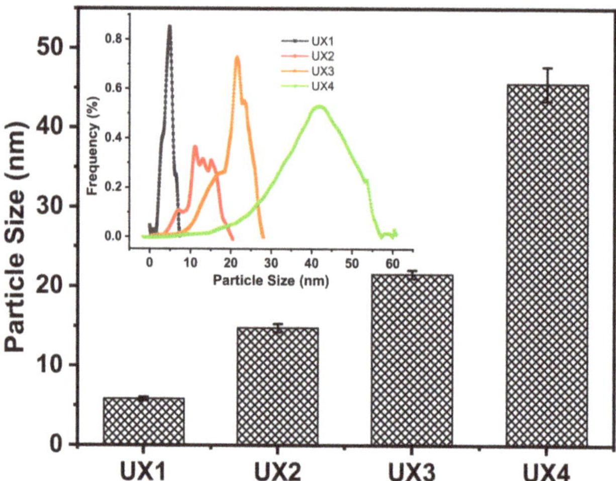

Figure 2. (a) Tyndall phenomenon, (b) AFM images, and (c) TEM images of xylan colloidal dispersions.

Figure 3. Particle size distribution and the average particle size of xylan colloidal dispersions.

3.3. Emulsifying Properties of XNCs

The xylan colloidal dispersions (UX1–UX4) were used as an emulsifying agent to produce oil—water emulsions and were compared with gum arabic (GA), Tween 80 (T80), and Triton X-100 (TX100). GA is a type of polysaccharide collected from *Acacia senegal* and is one of the most common natural food emulsifiers. T80 and TX100 are typical synthetic non-ionic surfactants. The obtained emulsions were tested for emulsifying activity (EA) and emulsion cream index (ECI), where EA refers to the ability of the emulsifier to form an emulsion and ECI corresponds to the stability of the formed emulsion [29]. For EA (Figure 4a–d), xylans had much better EA than GA at all emulsifier concentration levels. Xylan dispersions showed lower EA values than synthetic emulsifiers at low emulsifier concentrations, but UX3 showed the highest EA (1.07) and UX4 showed a comparable EA (0.84) to synthetic emulsifiers (0.92 for T80 and 0.84 for TX100) at a high emulsifier concentration of 4 wt%. For ECI (Figure 4e–h), in general, xylan dispersions showed emulsifying stabilities comparable to the synthetic emulsifiers, except at a very low emulsifier concentration of 0.5 wt%. UX3 in particular showed an excellent ECI that was maintained below 5% after centrifugation for 24 min. UX1 also showed an ECI maintained below 5% after 24 min, but only at a high emulsifier concentration of 4 wt%. The emulsifying stability of xylan dispersions can also be proved by the pictures of emulsions after standing for seven days (Figures S3 and S4). The particle size of the emulsion is also an important indicator for the evaluation of the formed emulsions [29]. Xylan-assisted emulsions showed larger and comparable particle sizes at emulsifier concentrations of 0.5% and 2%, respectively, compared to those of T80 and TX100, but they also showed narrower particle distributions than those of T80 and TX100 (Figure 5).

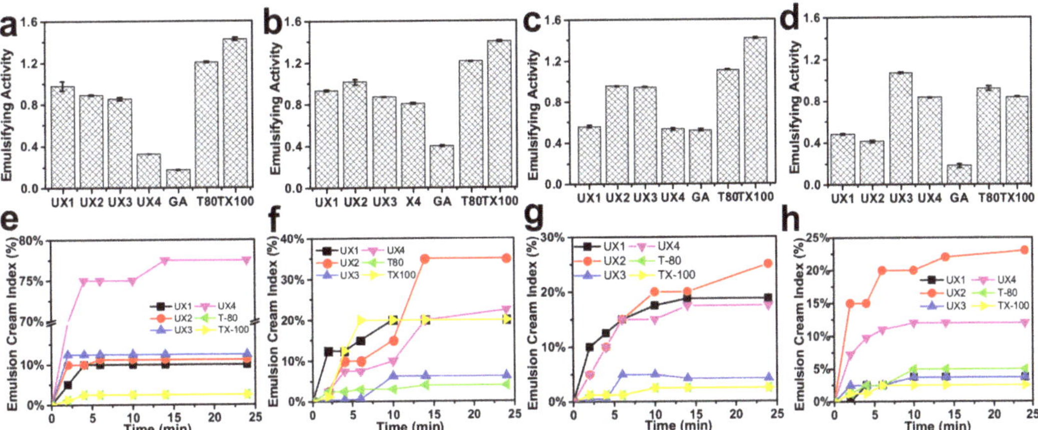

Figure 4. Emulsifying activity (**a–d**) and emulsion cream index after centrifugation (**e–h**) at 0.5, 1, 2, and 4 wt% emulsifier concentrations.

In general, xylan colloidal dispersions had much better emulsifying properties than those of GA and had comparable emulsifying properties to synthetic emulsifiers (Figure 4). T-80 and TX100 have a high surface activity as shown by their low surface tension (Figure 6a) and, therefore, can be quickly adsorbed onto the oil–water interface to assist the stabilization of an emulsion [30]. The emulsifying properties of xylan dispersions are quite dependent on their concentrations, which may be due to the fact that the viscosity of xylan dispersions is concentration dependent and shear thinning (Figure 6b,c). Previous research has shown that the rheological property of emulsifiers plays an important role in stabilizing emulsions, besides interfacial adsorption [31,32]. This suggests that UX1 has a good ECI, which is comparable to that of T80 and TX100 only at a high emulsifier concentration. This also explains the better emulsifying properties of UX3 than UX4, since UX3 has a higher viscosity than that of UX4 (Figures 4b–d and 6b,c).

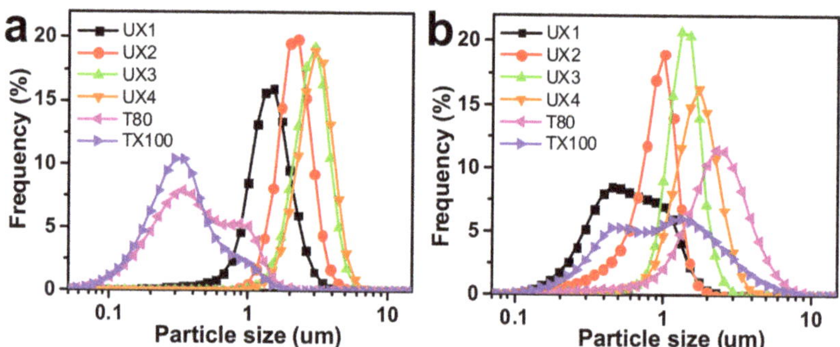

Figure 5. Particle size distribution of emulsions with (**a**) 0.5 wt% and (**b**) 2 wt% emulsifiers.

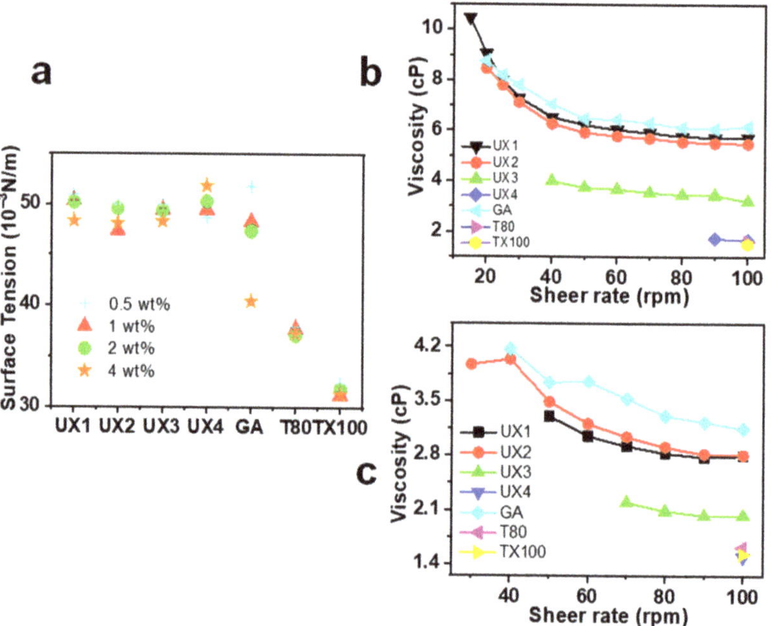

Figure 6. Surface tension (**a**), rheological viscosity of xylan colloidal dispersions, and other emulsifiers at 4 wt% (**b**) and 2 wt% (**c**) concentrations.

Comparing the different xylan colloidal dispersions, UX3 and UX4 had better emulsifying properties than UX1 and UX2, which may be due to the formation of uniformly sized XNC particles in UX3 and UX4 (Figure 2). The assistance of XNC particles in the emulsions can be better proved by dyeing the emulsifiers beforehand and observing their distribution in the emulsion by fluorescence [15]. As shown in Figure 7a, no obvious fluorescence was observed on the oil droplet surface of the UX1 and UX2 emulsions. On the contrary, an obvious green fluorescence appeared on the oil droplet surface with full coverage for the UX3 and UX4 emulsions. This suggested the forming of Pickering emulsions with the assistance of XNCs.

To further analyze the assistance of XNCs in Pickering emulsions, the oil phase was replaced with styrene, which was emulsified with water by UX3 and UX4. Styrene was polymerized in situ within its droplets, and rigid polystyrene microspheres were obtained (Figure 7b). The polystyrene microspheres showed a large size due to the avoidance of severe mechanical treatment during polymerization in the formed emulsion, which was conducive to observing the distribution of XNCs onto the surface of the microspheres [17,33]. By adding XNCs as the emulsifier, XNC particles adsorbed onto the surface of the polystyrene microspheres were clearly observed (Figure 7c,d). The dense coverage of XNCs on the oil (styrene) droplet surface formed a strong barrier layer to keep the styrene droplets in a spherical shape during the process of styrene polymerization. This again proves the high-density adsorption of XNCs on oil—water interface, preventing the aggregation of oil droplets and the forming of stable Pickering emulsions assisted by XNCs. In addition, UX3 emulsion droplets adsorbed more XNC particles on the surface than UX4 emulsion droplets (Figure 7c,d), which indicates that an XNC with a smaller size is more conducive to its adsorption onto the oil—water interface. This is another explanation of the better emulsifying properties of UX3 in comparison to UX4.

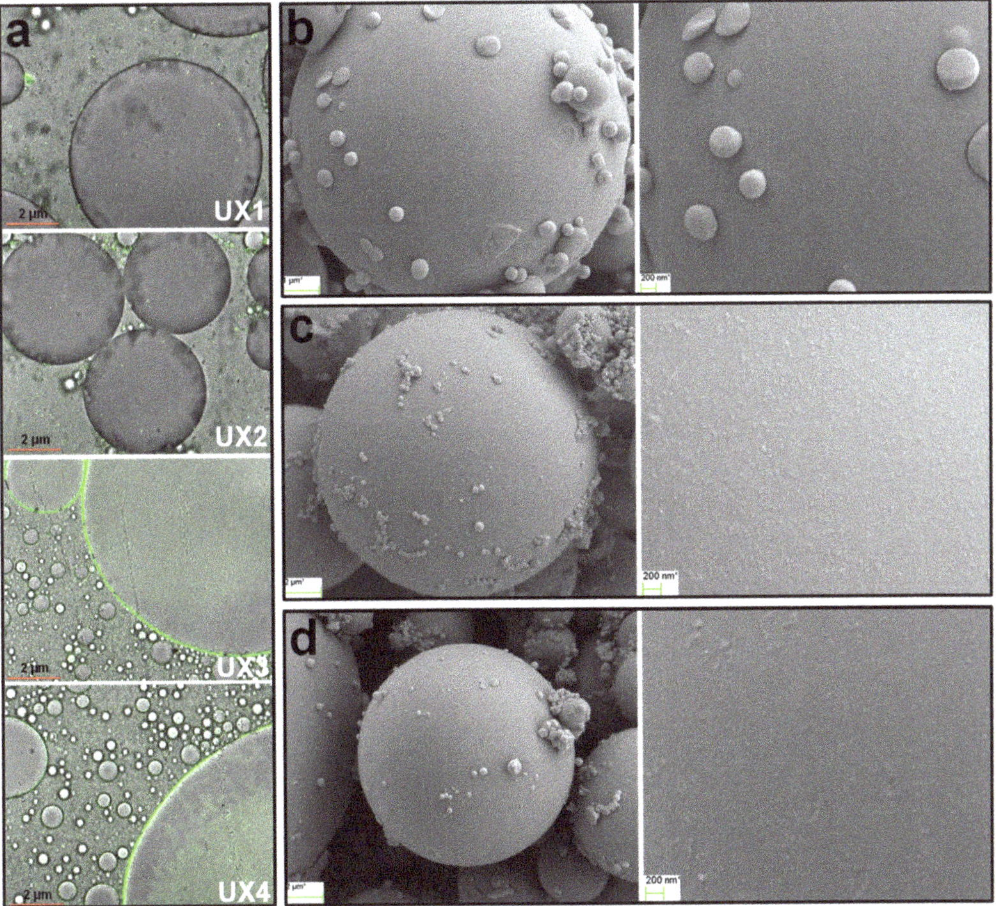

Figure 7. (**a**) Laser confocal microscope images of UX1–UX4 adsorption on the oil droplet surface. FESEM images of the rigid polystyrene droplet (produced from in situ polymerization) surface: (**b**) no emulsifier, (**c**) UX3, and (**d**) UX4 as the emulsifier.

Noncrystalline and soluble xylans are able to provide emulsification [32], but our study shows that, without the ability of forming Pickering emulsions, noncrystalline xylans, i.e., UX1 and UX2, demonstrate lower emulsifying properties than XNCs, i.e., UX3 and UX4. As shown in Figure 7a, UX1 and UX2 as emulsifiers have almost no oil—water interfacial adsorption in the emulsions. That is, the stability of UX1 and UX2 emulsions depends mainly on the gel network formed by the emulsifier, namely viscosity [31,32]. For UX3 and UX4 emulsions, with the assistance of XNC interfacial adsorption, their emulsion stability is greatly improved.

In comparison with some other common biopolymer nanocrystals, e.g., CNCs, chitin nanocrystals, and starch nanocrystals, UX3 and UX4 had better emulsifying properties (Table 4). In comparison with synthetic emulsifiers, e.g., T80 and TX100, UX3 and UX4 demonstrated comparable emulsifying properties. Considering that the production of XNCs is completely green without any acid treatments or chemical derivatizations, we state that XNCs have great potential to substitute synthetic chemicals in emulsification.

Table 4. Comparisons with other natural polysaccharide nanocrystals in the Pickering emulsion.

Emulsifier	Oil–Water Ratio	Stability Condition	* ECI (%)	Reference
CNC	1:4 (*w:w*)	Measured right after	$\geq$55	[17]
CNC with CNF	1:99 (*w:w*)	Stand for 7 days	95–0	[15]
CNC from bacterial cellulose	3:7	Centrifuged at 4000 g for 10 min	80–10	[33]
CNC	3:7	Centrifuged at 4000 g for 10 min	$\geq$70	[34]
Starch nanocrystals	1:1 (*v:v*)	Stand for 2 months	20–0	[21]
Starch nanocrystals	1:3 to 3:1	Stand for 24 h	84–10	[35]
Chitin nanocrystals	2:8	Measured right after	70–30	[16]
Chitin nanocrystals	2:8	Measured right after	60–10	[19]
XNC	2:8 (*w:w*)	Centrifuged at 4000 g for 10 min/stand for 7 days	Close to 0	This work

* Appropriate estimation according to the pictures or tables in the literature.

4. Conclusions

The formation of xylan hydrate crystals can be induced by regulating the degree of branching of xylans. Xylan hydrate nanocrystals (XNCs) with a uniform size can be produced by a mild ultrasonic treatment introduced to the suspensions of xylan crystals. Noncrystalline and soluble xylans stabilize oil-in-water emulsions by forming a gel network, which is viscosity dependent. XNCs have better oil-in-water emulsifying properties than noncrystalline and soluble xylans, since XNCs can be adsorbed onto oil–water interfaces at a high density to form stable Pickering emulsions. XNCs demonstrate comparable emulsifying properties compared to synthetic emulsifiers, e.g., T80 and TX100. Considering that the production of XNCs is completely green without any acid treatments or chemical derivatizations, these results suggest that XNCs have great potential to substitute synthetic chemicals in emulsification.

Supplementary Materials: The following are available online at https://www.mdpi.com/article/10.3390/nano11102558/s1, Figure S1: Polarizing microscope images of xylan components in 0.5 wt% aqueous solution, Figure S2: Dispersibility/solubility of xylans with different crystallinity, Figure S3: The emulsion cream index of X1-X4 after standing for seven days, Figure S4: Pictures of emulsion after standing for 7 days at different emulsifier concentration.

Author Contributions: Conceptualization, Z.X.; methodology, S.W.; formal analysis, S.W.; investigation, S.W.; writing—original draft preparation, S.W.; writing—review and editing, Z.X.; supervision, Z.X.; project administration, Z.X.; funding acquisition, Z.X. All authors have read and agreed to the published version of the manuscript.

Funding: This research was funded by the Natural Science Foundation of Guangdong Province (2020A1515011021).

Conflicts of Interest: The authors declare no conflict of interest.

References

1. Smith, P.J.; Wang, H.-T.; York, W.S.; Peña, M.J.; Urbanowicz, B.R. Designer biomass for next-generation biorefineries: Leveraging recent insights into xylan structure and biosynthesis. *Biotechnol. Biofuels* **2017**, *10*, 286–299. [CrossRef]
2. Ren, J.-L.; Sun, R.-C. Hemicelluloses. In *Cereal Straw as a Resource for Sustainable Biomaterials and Biofuels*; Sun, R.-C., Ed.; Elsevier: Amsterdam, The Netherlands, 2010; pp. 73–130. [CrossRef]
3. Ebringerová, A.; Heinze, T. Xylan and xylan derivatives–biopolymers with valuable properties, 1. Naturally occurring xylans structures, isolation procedures and properties. *Macromol. Rapid Commun.* **2000**, *21*, 542–556. [CrossRef]
4. Mazeau, K.; Moine, C.; Krausz, P.; Gloaguen, V. Conformational analysis of xylan chains. *Carbohydr. Res.* **2005**, *340*, 2752–2760. [CrossRef] [PubMed]
5. Nieduszynski, I.A.; Marchessault, R.H. Structure of β,D(1->4′)-xylan hydrate. *Biopolymers* **1972**, *11*, 1335–1344. [CrossRef]
6. Xiang, Z.; Jin, X.; Huang, C.; Li, L.; Wu, W.; Qi, H.; Nishiyama, Y. Water cast film formability of sugarcane bagasse xylans favored by side groups. *Cellulose* **2020**, *27*, 7307–7320. [CrossRef]
7. Meng, Z.; Sawada, D.; Laine, C.; Ogawa, Y.; Virtanen, T.; Nishiyama, Y.; Tammelin, T.; Kontturi, E. Bottom-up construction of xylan nanocrystals in dimethyl sulfoxide. *Biomacromolecules* **2021**, *22*, 898–906. [CrossRef]
8. Hansen, N.M.L.; Plackett, D. Sustainable films and coatings from hemicelluloses: A review. *Biomacromolecules* **2008**, *9*, 1493–1505. [CrossRef]
9. Li, Q.; Wang, S.; Jin, X.; Huang, C.; Xiang, Z. The application of polysaccharides and their Derivatives in pigment, barrier, and functional paper coatings. *Polymers* **2020**, *12*, 1837. [CrossRef]
10. Andrewartha, K.A.; Phillips, D.R.; Stone, B.A. Solution properties of wheat-flour arabinoxylans and enzymically modified arabinoxylans. *Carbohydr. Res.* **1979**, *77*, 191–204. [CrossRef]
11. Lin, N.; Huang, J.; Dufresne, A. Preparation, properties and applications of polysaccharide nanocrystals in advanced functional nanomaterials: A review. *Nanoscale* **2012**, *4*, 3274–3294. [CrossRef]
12. Abdullah; Weiss, J.; Ahmad, T.; Zhang, C.; Zhang, H. A review of recent progress on high internal-phase Pickering emulsions in food science. *Trends Food Sci. Technol.* **2020**, *106*, 91–103. [CrossRef]
13. Chanzy, H.; Comtat, J.; Dube, M.; Marchessault, R.H. Enzymatic degradation of β(1→4) xylan single crystals. *Biopolymers* **1979**, *18*, 2459–2464. [CrossRef]
14. Wang, S.; Song, T.; Qi, H.; Xiang, Z. Exceeding high concentration limits of aqueous dispersion of carbon nanotubes assisted by nanoscale xylan hydrate crystals. *Chem. Eng. J.* **2021**, *419*, 129602–129609. [CrossRef]
15. Bai, L.; Huan, S.; Xiang, W.; Rojas, O.J. Pickering emulsions by combining cellulose nanofibrils and nanocrystals: Phase behavior and depletion stabilization. *Green Chem.* **2018**, *20*, 1571–1582. [CrossRef]
16. Jimenez-Saelices, C.; Trongsatitkul, T.; Lourdin, D.; Capron, I. Chitin Pickering emulsion for oil inclusion in composite films. *Carbohydr. Polym.* **2020**, *242*, 116366–116375. [CrossRef]
17. Li, X.; Li, J.; Gong, J.; Kuang, Y.; Mo, L.; Song, T. Cellulose nanocrystals (CNCs) with different crystalline allomorph for oil in water Pickering emulsions. *Carbohydr. Polym.* **2018**, *183*, 303–310. [CrossRef] [PubMed]
18. Tang, C.; Spinney, S.; Shi, Z.; Tang, J.; Peng, B.; Luo, J.; Tam, K.C. Amphiphilic cellulose nanocrystals for enhanced Pickering emulsion stabilization. *Langmuir* **2018**, *34*, 12897–12905. [CrossRef] [PubMed]
19. Perrin, E.; Bizot, H.; Cathala, B.; Capron, I. Chitin nanocrystals for Pickering high internal phase emulsions. *Biomacromolecules* **2014**, *15*, 3766–3771. [CrossRef] [PubMed]
20. Lu, X.; Wang, Y.; Li, Y.; Huang, Q. Assembly of Pickering emulsions using milled starch particles with different amylose/amylopectin ratios. *Food Hydrocoll.* **2018**, *84*, 47–57. [CrossRef]
21. Azfaralariff, A.; Fazial, F.F.; Sontanosamy, R.S.; Nazar, M.F.; Lazim, A.M. Food-grade particle stabilized pickering emulsion using modified sago (*Metroxylon sagu*) starch nanocrystal. *J. Food Eng.* **2020**, *280*, 109973–109974. [CrossRef]
22. Kobayashi, K.; Kimura, S.; Heux, L.; Wada, M. Crystal transition between hydrate and anhydrous (1→3)-β-D-xylan from Penicillus dumetosus. *Carbohydr. Polym.* **2013**, *97*, 105–110. [CrossRef]
23. Jin, X.; Hu, Z.; Wu, S.; Song, T.; Yue, F.; Xiang, Z. Promoting the material properties of xylan-type hemicelluloses from the extraction step. *Carbohydr. Polym.* **2019**, *215*, 235–245. [CrossRef]
24. Xiang, Z.; Watson, J.; Tobimatsu, Y.; Runge, T. Film-forming polymers from distillers' grains: Structural and material properties. *Ind. Crop. Prod.* **2014**, *59*, 282–289. [CrossRef]
25. Filisetti-Cozzi, T.M.; Carpita, N.C. Measurement of uronic acids without interference from neutral sugars. *Anal. Biochem.* **1991**, *197*, 157–162. [CrossRef]
26. Hu, Z.; Xiang, Z.; Lu, F. Synthesis and emulsifying properties of long-chain succinic acid esters of glucuronoxylans. *Cellulose* **2019**, *26*, 3713–3724. [CrossRef]
27. Bosmans, T.J.; Stépán, A.M.; Toriz, G.; Renneckar, S.; Karabulut, E.; Wågberg, L.; Gatenholm, P. Assembly of debranched xylan from solution and on nanocellulosic surfaces. *Biomacromolecules* **2014**, *15*, 924–930. [CrossRef]
28. Zen, A.; Saphiannikova, M.; Neher, D.; Grenzer, J.; Grigorian, S.; Pietsch, U.; Asawapirom, U.; Janietz, S.; Scherf, U.; Lieberwirth, I.; et al. Effect of molecular weight on the structure and crystallinity of poly(3-hexylthiophene). *Macromolecules* **2006**, *39*, 2162–2171. [CrossRef]

29. Xiang, Z.; Runge, T. Emulsifying properties of succinylated arabinoxylan-protein gum produced from corn ethanol residuals. *Food Hydrocoll.* **2016**, *52*, 423–430. [CrossRef]
30. McClements, D.J.; Jafari, S.M. Improving emulsion formation, stability and performance using mixed emulsifiers: A review. *Adv. Colloid Interface Sci.* **2018**, *251*, 55–79. [CrossRef]
31. Lam, S.; Velikov, K.P.; Velev, O.D. Pickering stabilization of foams and emulsions with particles of biological origin. *Curr. Opin. Colloid Interface Sci.* **2014**, *19*, 490–500. [CrossRef]
32. Mikkonen, K.S. Strategies for structuring diverse emulsion systems by using wood lignocellulose-derived stabilizers. *Green Chem.* **2020**, *22*, 1019–1037. [CrossRef]
33. Kalashnikova, I.; Bizot, H.; Bertoncini, P.; Cathala, B.; Capron, I. Cellulosic nanorods of various aspect ratios for oil in water Pickering emulsions. *Soft Matter* **2013**, *9*, 952–959. [CrossRef]
34. Kalashnikova, I.; Bizot, H.; Cathala, B.; Capron, I. Modulation of cellulose nanocrystals amphiphilic properties to stabilize oil/water interface. *Biomacromolecules* **2012**, *13*, 267–275. [CrossRef]
35. Ahmad, A.; Lazim, A. Evaluation of emulsion emulsified by starch nanocrystal: A preliminary study. *AIP Conf. Proc.* **2018**, *1940*, 0200831–0200835.

 *nanomaterials*

Communication

Thermal Conductivity Analysis of Chitin and Deacetylated-Chitin Nanofiber Films under Dry Conditions

Jiahao Wang [1], Keitaro Kasuya [1], Hirotaka Koga [2], Masaya Nogi [2] and Kojiro Uetani [2,*]

[1] Graduate School of Engineering, Osaka University, Mihogaoka 8-1, Ibaraki-shi, Osaka 567-0047, Japan; w541725924@eco.sanken.osaka-u.ac.jp (J.W.); keitaro_k@chem.eng.osaka-u.ac.jp (K.K.)

[2] The Institute of Scientific and Industrial Research (SANKEN), Osaka University, Mihogaoka 8-1, Ibaraki-shi, Osaka 567-0047, Japan; hkoga@eco.sanken.osaka-u.ac.jp (H.K.); nogi@eco.sanken.osaka-u.ac.jp (M.N.)

* Correspondence: uetani@eco.sanken.osaka-u.ac.jp; Tel.: +81-6-6879-8442

Abstract: Chitin, a natural polysaccharide polymer, forms highly crystalline nanofibers and is expected to have sophisticated engineering applications. In particular, for development of next-generation heat-transfer and heat-insulating materials, analysis of the thermal conductivity is important, but the thermal conductivity properties of chitin nanofiber materials have not been reported. The thermal conductivity properties of chitin nanofiber materials are difficult to elucidate without excluding the effect of adsorbed water and analyzing the influence of surface amino groups. In this study, we aimed to accurately evaluate the thermal conductivity properties of chitin nanofiber films by changing the content of surface amino groups and measuring the thermal diffusivity under dry conditions. Chitin and deacetylated-chitin nanofiber films with surface deacetylation of 5.8% and 25.1% showed in-plane thermal conductivity of 0.82 and 0.73 W/mK, respectively. Taking into account that the films had similar crystalline structures and almost the same moisture contents, the difference in the thermal conductivity was concluded to only depend on the amino group content on the fiber surfaces. Our methodology for measuring the thermal diffusivity under conditioned humidity will pave the way for more accurate analysis of the thermal conductivity performance of hydrophilic materials.

Keywords: chitin nanofiber; deacetylated chitin; nanopaper; thermal diffusivity

Citation: Wang, J.; Kasuya, K.; Koga, H.; Nogi, M.; Uetani, K. Thermal Conductivity Analysis of Chitin and Deacetylated-Chitin Nanofiber Films under Dry Conditions. *Nanomaterials* **2021**, *11*, 658. https://doi.org/10.3390/nano11030658

Academic Editor: Luis Morellón

Received: 21 February 2021
Accepted: 5 March 2021
Published: 8 March 2021

Publisher's Note: MDPI stays neutral with regard to jurisdictional claims in published maps and institutional affiliations.

1. Introduction

Chitin is a linear crystalline polymer consisting of N-acetylglucosamine with β-(1→4) linkages that is widely produced by insects [1], fungi [2,3], and crustaceans [4]. The crystalline chitin nanofibers extracted from native chitins have uses as engineering materials, such as high-strength composites [5], low-thermal expansion and transparent films [6,7], electronics [8], microparticles [9], membranes [10], and chiral nematic mesoporous materials [11], as well as in biomedical applications [12]. In particular, chitin nanofibers have been reported to have extremely high strength owing to their high crystallinity [13], and there are high expectations for their use in engineering [14,15]. Recently, it has been reported that highly crystalline cellulose nanofibers have high thermal conductivity compared with plastics and glass [16], and unique thermal application pathways have been developed, such as transparent and thermal conductive films [17], directional heat flow films [18,19], and heat-transfer modulation films [20]. Nanofibers of chitin, a similar polysaccharide to cellulose with high crystallinity, are expected to have similar properties, but thermal conductivity analysis of chitin nanofiber materials has not been performed.

Clarification of the thermal conductivity properties of chitin nanofiber materials toward unexplored thermal applications remains a major challenge. In analyzing the thermal conductivity properties of chitin nanofiber materials, there are three inherent problems. First, natural chitin has a small content of amino groups (chitosan segments), which results in an irregular chemical structure on the nanofiber surface. It is, therefore, important to evaluate the effect of the different content of amino groups on the thermal conductivity. Second, chitin nanofibers have a highly anisotropic fibrous morphology, and when they are dried as a nonwoven film, there are large structural differences between the film's in-plane and through-plane directions. It is necessary to separately measure the thermal conductivity properties for each direction. Third, because chitin nanofibers are hydrophilic, like cellulose, moisture adsorption will occur when the nanofiber films are investigated in the ambient environment. This moisture adsorption is known to have a strong tendency to swell and irregularly bend the films. To accurately analyze the thermal conductivity of the chitin nanofiber film, it is necessary to measure the thermal conductivity while keeping the film as dry as possible.

In this study, we aimed to elucidate the thermal conductivity properties of chitin nanofiber films by technically addressing the above three issues. For the first issue, partially deacetylated samples were prepared by alkaline treatment to clearly increase the content of surface amino groups, according to a previous study [21]. For the second issue, we used the nondestructive and noncontact method of the laser spot periodic heating radiation thermometry method [22], which allows independent measurement of film-like materials in the in-plane and through-plane directions. For the third issue, a specially designed sealed chamber was used to realize the thermal diffusivity measurement in a very low humidity environment.

2. Materials and Methods

2.1. Materials

A total of 175 g of the leg shell of snow crab (*Chionoecetes opilio*) was boiled for 1.5 h and then cut into ~2 cm × 2 cm square pieces. The shells were then immersed in 2 L of 5 wt% NaOH for 6 h at 90 °C without vigorous stirring. Subsequently, the product was cooled to room temperature and filtered to repeatedly wash with distilled water until the pH reached neutral. The product was then immersed in 2 L of 5 wt% HCl at room temperature for 48 h, followed by filtering to remove minerals and proteins. These alkaline and acid treatments were repeated another two times. For these repetitions, the treatment time was reduced to 2 h. The product was then bleached with 1.5 L of aqueous $NaClO_2$ solution at 80–90 °C under acidic conditions, where the pH was adjusted to ~3 by adding acetic acid. We added 15 g of $NaClO_2$ and acetic acid per hour for 3 h, followed by repeating for another 3 h with a decreased $NaClO_2$ amount of 7.5 g per hour. After complete bleaching, the sample was washed to obtain bleached crab shell.

Wet bleached crab shell (13.6 g, corresponding to dry weight of 3 g) was suspended in 75 mL of 33.6 wt% NaOH solution and mixed with 0.09 g $NaBH_4$ at 90 °C for 4 h [21]. $NaBH_4$ was added to prevent depolymerization [21,23]. During the reaction, the beaker was shaken every 15–20 min. The product was then repeatedly washed with distilled water to obtain deacetylated crab shell.

The bleached and deacetylated crab shells were used to produce chitin nanofiber (ChNF) and deacetylated-ChNF (D-ChNF), respectively. Both of the crab shells dispersed in 500 mL of distilled water were agitated by a high-speed blender with a 2 L stainless-steel bottle (CAC90B X-TREME, Waring Commercial, Conair Corp., Stamford, CT, USA) combined with an Absolute3 motor (Vitamix, Cleveland, OH, USA) at 24,000 rpm for 12 min for preliminary fibrillation. The suspensions diluted to 1 L were then passed through a high-pressure water-jet system (Starburst 10, Sugino Machine Limited, Toyama, Japan) with a ball-type chamber at 180 MPa for 50 cycles for fibrillation into nanofibers. The final suspensions for ChNF and D-ChNF had concentrations of 0.20 and 0.16 wt%, respectively.

2.2. Film Fabrication

The ChNF and D-ChNF suspensions were filtered by a membrane filter (A010A047A, Advantec Toyo Kaisha, Ltd., Tokyo, Japan) to form wet mats. Another membrane filter was attached to the upper surface of the wet mats, and they were then hot-pressed at 110 °C for 15 min to dry the nanofiber films with thickness of ~55 μm.

2.3. Degree of Deacetylation

The degree of deacetylation of the nanofiber samples was measured through conductometric titration. The suspensions of the never-dried nanofiber samples were further diluted to 0.12–0.17 wt% with distilled water, and 5 mL of 0.01 M NaCl solution was added. The pH was then adjusted to ~10 by adding 0.05 M NaOH. The suspension was stirred for 30 min, followed by adjusting the pH to ~2.8 by adding 0.1 M HCl. The suspension was then titrated by adding 0.05 M NaOH at a speed of 0.1 mL/min using an automatic titrator (AUT-701, DKK-TOA Corp., Tokyo, Japan).

According to a previous method [24], the degree of deacetylation was calculated using the following equations:

$$n \ (\text{mmol/g}) = \frac{V \times c}{w} \tag{1}$$

$$\text{Degree of deacetylation} \ (\%) = \frac{203.1927 \times n/1000}{1 + 42.0358 \times n/1000} \tag{2}$$

where n, V, c, and w are the amino group content, consumed volume of NaOH solution at the plateau region of the titration profile, concentration of the titrated NaOH, and weight of the measured sample, respectively.

2.4. Characterization

X-ray diffractometry was performed on the films prepared in Section 2.2 using a Miniflex600 diffractometer (Rigaku Corp., Tokyo, Japan) with monochromatic Cu Kα radiation generated at 40 kV and 15 mA. The scattering angle 2θ was scanned from 5° to 35° at steps of 0.01°. The crystallite width (D) of each diffraction plane (*hkl*) was calculated by the method of full width at half-maximum (fwhm) using Scherrer's equation: $D_{hkl} = K\lambda/\beta \cos\theta$, where $K = 0.9$, λ is the wavelength (1.541862 Å), β is the fwhm of the peak, and θ is the Bragg angle. The relative degree of crystallinity I_C was estimated from the area ratios of the crystalline peaks to the sum of the separated peaks.

Field emission scanning electron microscopy (FESEM, JSM-F100, JEOL Ltd., Tokyo, Japan) observation was performed at an acceleration voltage of 1 kV. The nanofiber films were prepared by casting 5 mL of the suspensions mixed with the same volume of acetone on a Teflon dish, drying in an oven at 55 °C, and then coating with Au by an ion sputterer (E-1045, Hitachi High-Tech Corp., Tokyo, Japan).

Transmission electron microscopy (TEM) observation was performed on two types of nanofibers using a JEM-ARM200F (JEOL, Ltd., Tokyo, Japan) at an acceleration voltage of 200 kV. Samples were negatively stained with the phosphotungstic acid.

The moisture content of the films conditioned under humidity below 10% at 25 °C for more than 1 day was calculated from the weight loss of the film heated at 100 °C for 60 min. After conditioning the film under constant humidity, the film was promptly introduced into a thermogravimetric analyzer (Q50, TA Instruments Japan, Tokyo, Japan) and the temperature was rapidly increased to 100 °C to perform isothermal gravimetry for 60 min under a nitrogen atmosphere. The measurements were performed three times per film.

2.5. Thermal Conductivity Measurement under Dry Conditions

The thermal conductivity of the film κ was calculated by the equation $\kappa = \alpha \rho C_\text{p}$, where α is the thermal diffusivity, which was measured by the unsteady method, C_p is the specific heat capacity, and ρ is the bulk density. In the following, the subscripts T and I for κ and α represent the through-plane and in-plane direction, respectively.

The thermal diffusivity was measured by the laser spot periodic heating radiation thermometry method using a TA33 thermowave analyzer (Bethel Co., Ltd., Ibaraki, Japan). Both sides of the film were blackened with graphite spray to prevent permeation of the laser and to keep the emissivity constant in order to ensure the accuracy of temperature detection. A sealing chamber was custom-made to measure the thermal diffusivity of the films under dry conditions. The sample was placed on a stainless-steel washer and then affixed to the sealing with polyimide tape. The sealing chamber holding the sample was conditioned to relative humidity of below 10% in the tightbox with an excess amount of dried silica gel for at least 1 day. After conditioning, the chamber lid was promptly tightened to seal the chamber under the conditioning environment, and the thermal diffusivity was measured. For measurement of the thermal diffusivity in the through-plane direction (α_T), the frequency of the heating laser f was set from 10.13 to 60.13 Hz with a step size of 2.0 Hz, and its intensity was fixed at 8%. Conversely, during detection of the thermal diffusivity in the in-plane direction (α_I), f was set to 0.16, 0.21, 0.31, 0.41, 0.51, and 0.61 Hz with the intensity of the heating laser set to 12%, and the distance was changed from 0.4 to 0.7 mm, with a step size of 0.03 mm.

The specific heat capacity C_p was measured by a differential scanning calorimeter (Thermo Plus Evo2 8230, Rigaku, Japan). The sample weight ranged from 4 to 6 mg and the scanning temperature ranged from -30 to 130 °C at a heating rate of 5 °C/min. We measured the second or third heating scan to remove the effect of absorbed moisture [16]. In brief, after heating the sample to 130 °C in the first scan, the sample temperature decreased to 40 °C at a rate of -30 °C/min, and the weight of the sample was remeasured. The sample was then remounted to cool to -30 °C for the second scan. C_p was calculated from the data of the second or third scan.

The bulk density of the nanofiber film was calculated from its weight and volume. The films were stored in a tightbox in the dry state (relative humidity below 10%) until just before the thickness and weight measurement.

3. Results

3.1. Characterization of the Nanofiber Samples

Both the ChNF and D-ChNF films showed a translucent appearance owing to their fine nanofiber network structures (Figure 1a,b). The apparent transparency is thought to be related to the chemical composition of the nanofiber surface and bulk density, where the bulk densities of the ChNF and D-ChNF films were 0.98 and 1.20 g/cm^3, respectively. In addition, the ChNF film might have slightly thicker bundles than D-ChNF, even after high-pressure water-jet treatment, which may also explain the smaller packing density and hazy appearance of the ChNF film. High resolution TEM observation (Figure 1c,d) shows that the ChNF tends to be slightly bunched and aggregated, whereas the D-ChNF tends to be well dispersed. Furthermore, the D-ChNF was found to have a short rod-like (needle-like) whisker shape. It was suggested that the strong alkali treatment for deacetylation and the subsequent water-jet fibrillation treatment shortened the fiber length.

The X-ray diffraction profiles of both the ChNF and D-ChNF samples (Figure 1e) showed typical α-chitin crystals [25,26]. After peak deconvolution, we found similar crystallite sizes and degrees of crystallinity for ChNF and D-ChNF, as displayed in Table 1, indicating that the α-chitin crystal structure was retained after harsh alkaline treatment using 33.6 wt% NaOH for 4 h, and only the nanofiber surface was deacetylated. This result is in good agreement with previous reports [21].

The conductometric titration profiles of ChNF and D-ChNF greatly differed (Figure 1f). For the change in the conductivity with titration of 0.05 M NaOH, the first slope at small titrating volumes was because of neutralization of the strong acid (hydrochloric acid), the plateau region in the center was because of neutralization of the targeted weak acid (amino group), and the following ascending slope was because of the increase in the concentration of the alkali. D-ChNF showed a significantly larger plateau region than ChNF, indicating that NaOH consumption by neutralization of amino groups was higher.

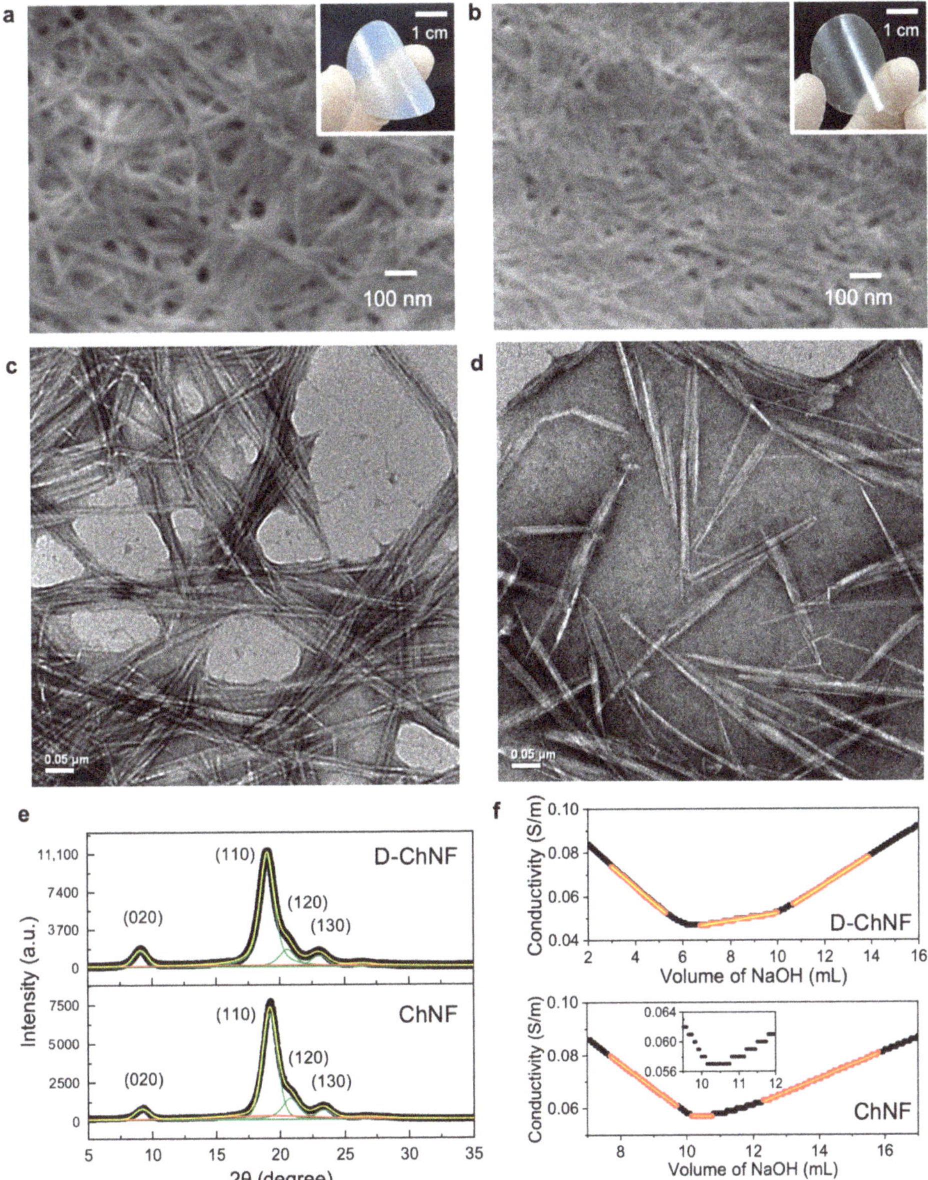

Figure 1. Characterization of D-ChNF and ChNF. FESEM images of (**a**) ChNF and (**b**) D-ChNF. The inserts show the appearance of the films. TEM images of (**c**) ChNF and (**d**) D-ChNF. (**e**) Typical X-ray diffraction profiles of D-ChNF and ChNF. The profiles were deconvoluted into the green and red peaks originating from crystalline and amorphous diffraction, respectively, and fitted by the cumulative fit (yellow profile). (**f**) Conductometric titration plots of D-ChNF and ChNF. The red plots were fitted with the yellow lines, and consumption of 0.05 M NaOH solution was calculated from the two intersections of the three fitted lines. The inset for ChNF shows the magnified plot at around the plateau region.

To quantify the content of amino groups, the three regions (Figure 1f) were linearly fitted, and the volume of 0.05 M NaOH solution consumed in the plateau region was calculated from the two intersections. The obtained NaOH volume gave the amino group content and degree of deacetylation (Table 2). ChNF contained a small content of inherent amino groups on its surfaces, and D-ChNF contained ~4.5 times greater content of amino groups than ChNF. These results were consistent with previous reports [21,27]. Thus, we succeeded in preparing ChNFs with different surface amino group contents toward investigating the effect of the amino group content on the thermal conductivity properties.

Table 1. Crystallographic analysis.

	d_{020}	d_{110}	d_{120}	d_{130}	I_C
	nm	nm	nm	nm	%
ChNF	5.87 ± 0.84	5.86 ± 0.11	5.42 ± 1.48	5.67 ± 0.30	83.17
D-ChNF	6.10 ± 0.29	5.85 ± 0.06	5.48 ± 0.42	5.47 ± 0.56	85.54

Table 2. Titration analysis to determine the degree of deacetylation.

	$-NH_2$ Content	Degree of Deacetylation
	mmol/g	%
ChNF	0.29	5.8
D-ChNF	1.30	25.1

3.2. Analysis of the Thermal Conductivity Properties under Dry Conditions

To measure the thermal diffusivity under dry conditions, we used a custom-made sealing chamber that fits the TA33 stage to keep the sample under dry conditions. The samples were fixed in the acrylic sealing chamber shown in Figure 2a and conditioned in a tightbox containing an excess amount of dried silica gel for more than 1 day with the lid open (Figure 2b). The relative humidity in the tightbox was kept below 10%. After sufficient conditioning time, the lid of the sealing chamber was quickly closed, and the thermal diffusivity was measured while maintaining the inside of the chamber under dry conditions (Figure 2c).

The validity of the thermal diffusivity measurements using the chamber was confirmed through measuring a standard sample of a 300-μm-thick copper plate and a cellulose nanofiber film. For all of the samples, the thermal diffusivity values measured with and without the chamber were comparable, and their variation was within the 5% measurement error of the TA33 thermowave analyzer. We concluded that it is possible to accurately measure the thermal diffusivity with a sufficiently small error range using the chamber.

The thermal diffusivities of the ChNF and D-ChNF films were measured under dry conditions. The humidity in the sealed chamber during the measurement was kept constant at 2–3%, as shown in Figure 3, confirming that the measurement was carried out under stable dry conditions owing to sufficient sealing. We waited for 5–15 min before starting the diffusivity measurement to stabilize the humidity inside the chamber.

The measured thermal diffusivity results are shown in Figure 4. In the measurement of the thermal diffusivity in the through-plane direction, the frequency of the periodic heating laser was varied at the same point of heating, and the thermal diffusivity was calculated from the rate of change of the phase measured on the sample backside (slope of the plot). For the in-plane direction, in addition to the heating frequency, the phase detection point on the sample backside needed to be changed, and the diffusivity was calculated from the rate of change of the phase with respect to the distance from the heating point [16,22].

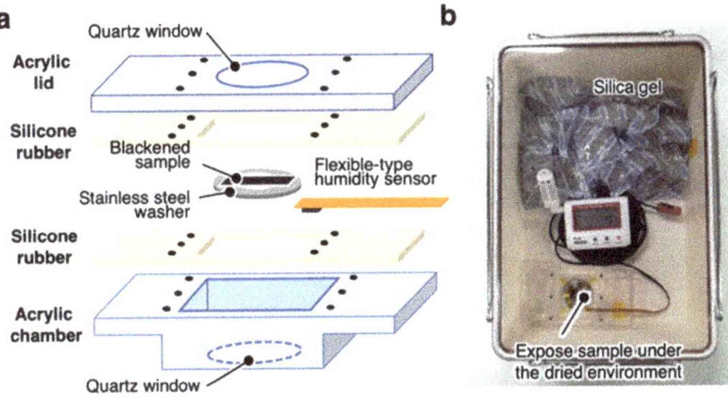

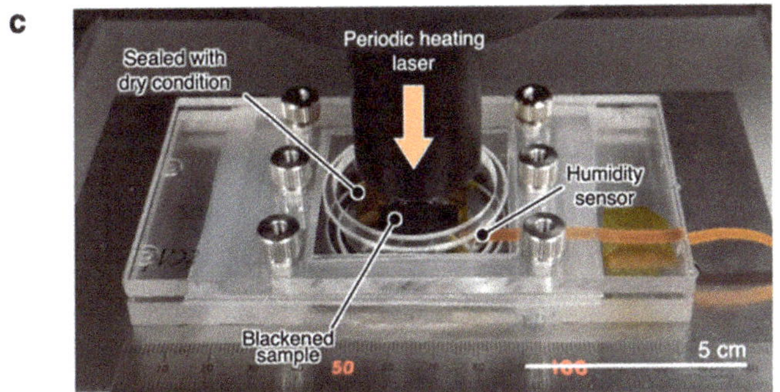

Figure 2. System for measuring the thermal diffusivity under dry conditions. (**a**) Setup of the sealing chamber. (**b**) Chamber holding the sample conditioned in the tightbox under relative humidity below 10% with an excess amount of dried silica gel. (**c**) Sealed chamber under dry conditions mounted on the TA33 stage to measure the thermal diffusivity.

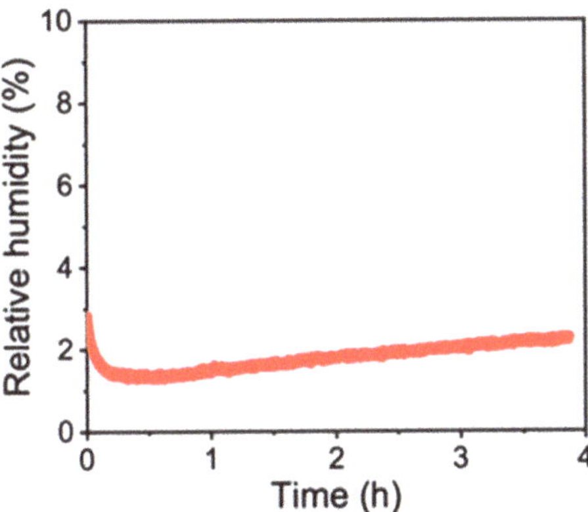

Figure 3. Typical change in the relative humidity in the sealed chamber during thermal diffusivity measurement.

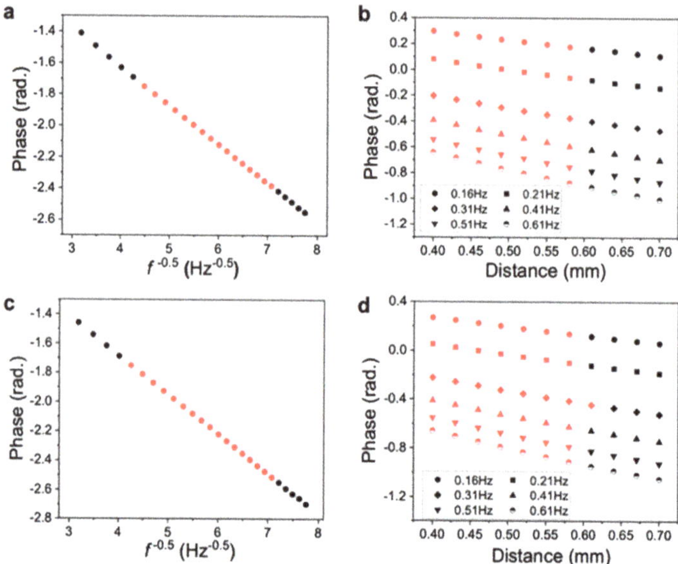

Figure 4. Typical experimental data for measuring the thermal diffusivity. Measurement data of the ChNF film in the (**a**) through-plane and (**b**) in-plane directions. Measurement data of the D-ChNF film in the (**c**) through-plane and (**d**) in-plane directions. The red points were used for linear fitting to determine the slopes.

The phase and amplitude of both ChNF and D-ChNF, indicating propagation of the temperature waves, were clearly detected and showed an ideal linear trend (Figure 4), confirming that the measurements were correctly performed. The calculated diffusivities are listed in Table 3. For both films, the thermal conductivity and thermal diffusivity in the through-plane direction were smaller than those in the in-plane direction. This is because of the fiber orientation of the layers within the nonwoven film [16,18]. The in-plane thermal conductivities of both films were in a similar range to those of wood cellulose nanofiber films [16].

Table 3. Thermal conductivity properties of the ChNF and D-ChNF films.

	ρ	C_p	α_T	α_I	κ_T	κ_I
	g/cm³	**J/gK**	**mm²/s**	**mm²/s**	**W/mK**	**W/mK**
ChNF film	0.98 ± 0.01	1.33 ± 0.06	0.152 ± 0.05	0.625 ± 0.06	0.22 ± 0.06	0.82 ± 0.12
D-ChNF film	1.20 ± 0.03	1.21 ± 0.01	0.135 ± 0.02	0.475 ± 0.01	0.20 ± 0.03	0.73 ± 0.01

While there were no significant differences in the thermal diffusivity and conductivity of the ChNF and D-ChNF films in the through-plane direction, the in-plane thermal diffusivity (α_I) of the D-ChNF film was about 25% lower than that of the ChNF film. The in-plane thermal conductivity of the D-ChNF film was also ~10% lower. The measurement error of the TA33 in measuring the thermal diffusivity is guaranteed to be within ±5%, and we considered the α_I difference between the films to be significant. This demonstrates that surface deacetylation of ChNF significantly increased the phonon scattering at the nanofiber interface to decrease thermal propagation within the films.

To investigate the cause of the thermal conductivity difference, the adsorption moisture content of the films was measured. The ChNF and D-ChNF films showed almost the same moisture contents of 3.21% ± 0.31% and 3.32% ± 0.08%, respectively (Figure 5). Thus, even though the contents of surface amino groups differed, the moisture content could be reduced by conditioning the films for a long time in a dry environment with a relative humidity of less than 10%. The dehydration behavior in the thermogravimetric measurements (Figure 5) showed that the moisture desorption rate of D-ChNF was lower than that of ChNF. This appears to be because of the higher surface hydrophilicity of D-ChNF, which is in good agreement with the trend in the content of amino groups. Overall, the difference in the thermal diffusivity detected in this study was determined not by the moisture content or crystal structure, but by the content of surface amino groups.

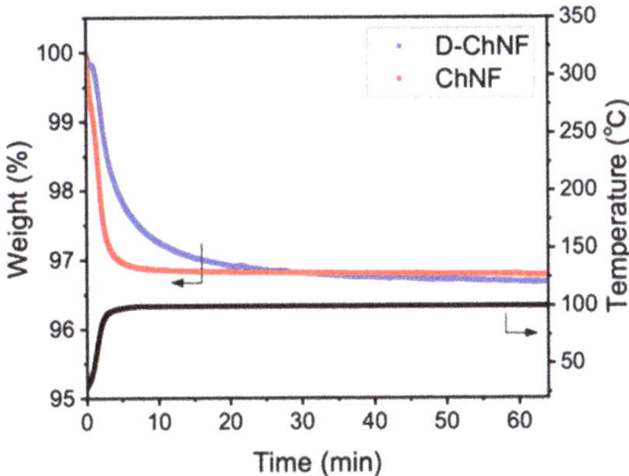

Figure 5. Typical profiles of thermogravimetric analysis of the ChNF and D-ChNF films just after conditioning at relative humidity below 10%.

4. Discussion

For nanofiber films of cellulose, a crystalline polysaccharide similar to chitin, the in-plane thermal conductivity is proportional to the crystallite width [16]. In particular, bacterial cellulose nanofiber films with similar crystallite widths to the ChNF film (approximately 5–6 nm) showed in-plane thermal conductivity of about 1.3 W/mK, although the drying conditions were not controlled [16], while the ChNF film showed significantly lower conductivity of 0.82 W/mK. The phonon propagation properties of crystalline nanofiber films were found to differ depending on their molecular composition, as well as the crystallite width.

The thermal diffusivity and thermal conductivity of the D-ChNF film, whose crystal structure and moisture content were almost the same as those of the ChNF film, but containing a larger content of surface amino groups, were clearly reduced compared with the ChNF film. Although the cationic charge increased with the increased content of amino groups, its introduction was partial (only 25%), and the amino groups were thought to be randomly located, because alkaline treatment did not have any site specificity, which may have resulted in an increase of the thermal resistance at the interfiber interface. Since a stronger deacetylation treatment would dissolve chitin, we used the NaOH concentration of ~33 wt%, which is reported not to dissolve chitin [21]. The nanofiber length suggested by the TEM observation is likely to have little effect on the thermal conductivity, according to the previous report [16]. The bulk density of the film is inversely proportional to the in-plane thermal diffusivity [20], so it is reasonable that the D-ChNF film in this study had a higher bulk density and lower in-plane thermal diffusivity than the ChNF film. Although it has been reported that the thermal conductivity of cellulose nanofiber films is modulated by physical perturbation of the fiber–fiber interface [20], further detailed investigation is necessary to enhance the thermal conductivity by chemical modification of nanofiber surfaces.

5. Conclusions

In this study, the thermal conductivity of hygroscopic ChNF films was successfully measured under dry conditions. The ChNF and D-ChNF films with surface deacetylation of 5.8% and 25.1% showed in-plane thermal conductivity of 0.82 and 0.73 W/mK, respectively. The films had similar crystalline structures and almost the same moisture content, so the difference in the thermal conductivity was concluded to only depend on the content of amino groups on the fiber surfaces. Our methodology for measuring the thermal diffusivity under conditioned humidity paves the way for more accurate analysis of the thermal conductivity performance of hydrophilic materials.

Author Contributions: Conceptualization, K.U.; methodology, K.U. and H.K.; formal analysis, J.W. K.K. and K.U.; investigation, J.W.; resources, J.W. and K.K.; data curation, J.W., K.K., H.K. and K.U.; writing—original draft preparation, K.U. and H.K.; writing—review and editing, H.K. and M.N.; supervision, M.N.; project administration, K.U.; funding acquisition, K.U. and M.N. All authors have read and agreed to the published version of the manuscript.

Funding: This work was partially supported by funding from the Japan Science and Technology Mirai Research and Development Program of the Japan Science and Technology Agency (Grant No. JPMJMI17ED), Ogasawara Toshiaki Memorial Foundation, Konica Minolta Science and Technology Foundation, and Kurita Water and Environment Foundation.

Acknowledgments: J.W. thanks Kinya Katano, Co-creation Bureau, Osaka University, for help with the DSC measurement and Yosuke Murakami, Comprehensive Analysis Center, Osaka University, for his kind assistance with FESEM and TEM observation. We thank Tim Cooper, PhD, from Edanz Group for editing a draft of this manuscript.

Conflicts of Interest: The authors declare no conflict of interest.

References

1. Sajomsang, W.; Gonil, P. Preparation and characterization of α-chitin from cicada sloughs. *Mater. Sci. Eng. C* **2010**, *30*, 357–363. [CrossRef]
2. Lenardon, M.D.; Munro, C.A.; Gow, N.A. Chitin synthesis and fungal pathogenesis. *Curr. Opin. Microbiol.* **2010**, *13*, 416–423. [CrossRef]
3. Jones, M.; Kujundzic, M.; John, S.; Bismarck, A. Crab vs. Mushroom: A Review of Crustacean and Fungal Chitin in Wound Treatment. *Mar. Drugs* **2020**, *18*, 64. [CrossRef]
4. Arbia, W.; Arbia, L.; Adour, L.; Amrane, A. Chitin Extraction from Crustacean Shells Using Biological Methods—A Review. *Food Technol. Biotechnol.* **2013**, *51*, 12–25.
5. Jin, J.; Hassanzadeh, P.; Perotto, G.; Sun, W.; Brenckle, M.A.; Kaplan, D.; Omenetto, F.G.; Rolandi, M. A biomimetic composite from solution self-assembly of chitin nanofibers in a silk fibroin matrix. *Adv. Mater.* **2013**, *25*, 4482–4487. [CrossRef] [PubMed]
6. Ifuku, S.; Morooka, S.; Morimoto, M.; Saimoto, H. Acetylation of Chitin Nanofibers and their Transparent Nanocomposite Films. *Biomacromolecules* **2010**, *11*, 1326–1330. [CrossRef] [PubMed]
7. Shams, M.I.; Ifuku, S.; Nogi, M.; Oku, T.; Yano, H. Fabrication of optically transparent chitin nanocomposites. *Appl. Phys. A* **2011**, *102*, 325–331. [CrossRef]
8. Jin, J.; Lee, D.; Im, H.G.; Han, Y.C.; Jeong, E.G.; Rolandi, M.; Choi, K.C.; Bae, B.S. Chitin Nanofiber Transparent Paper for Flexible Green Electronics. *Adv. Mater.* **2016**, *28*, 5169–5175. [CrossRef]
9. Kaku, Y.; Fujisawa, S.; Saito, T.; Isogai, A. Synthesis of Chitin Nanofiber-Coated Polymer Microparticles via Pickering Emulsion. *Biomacromolecules* **2020**, *21*, 1886–1891. [CrossRef]
10. Mushi, N.E.; Butchosa, N.; Salajkova, M.; Zhou, Q.; Berglund, L.A. Nanostructured membranes based on native chitin nanofibers prepared by mild process. *Carbohydr. Polym.* **2014**, *112*, 255–263. [CrossRef] [PubMed]
11. Nguyen, T.-D.; MacLachlan, M.J. Biomimetic Chiral Nematic Mesoporous Materials from Crab Cuticles. *Adv. Optical Mater.* **2014**, *2*, 1031–1037. [CrossRef]
12. Ifuku, S.; Tsukiyama, Y.; Yukawa, T.; Egusa, M.; Kaminaka, H.; Izawa, H.; Morimoto, M.; Saimoto, H. Facile preparation of silver nanoparticles immobilized on chitin nanofiber surfaces to endow antifungal activities. *Carbohydr. Polym.* **2015**, *117*, 813–817. [CrossRef]
13. Bamba, Y.; Ogawa, Y.; Saito, T.; Berglund, L.A.; Isogai, A. Estimating the Strength of Single Chitin Nanofibrils via Sonication-Induced Fragmentation. *Biomacromolecules* **2017**, *18*, 4405–4410. [CrossRef]
14. Rolandi, M.; Rolandi, R. Self-assembled chitin nanofibers and applications. *Adv. Colloid Interface Sci.* **2014**, *207*, 216–222. [CrossRef] [PubMed]

15. Mushi, N.E.; Utsel, S.; Berglund, L.A. Nanostructured biocomposite films of high toughness based on native chitin nanofibers and chitosan. *Front. Chem.* **2014**, *18*, 99. [CrossRef]
16. Uetani, K.; Okada, T.; Oyama, H.T. Crystallite Size Effect on Thermal Conductive Properties of Nonwoven Nanocellulose Sheets. *Biomacromolecules* **2015**, *16*, 2220–2227. [CrossRef] [PubMed]
17. Uetani, K.; Okada, T.; Oyama, H.T. Thermally conductive and optically transparent flexible films with surface-exposed nanocellulose skeletons. *J. Mater. Chem. C* **2016**, *4*, 9697–9703. [CrossRef]
18. Uetani, K.; Okada, T.; Oyama, H.T. In-Plane Anisotropic Thermally Conductive Nanopapers by Drawing Bacterial Cellulose Hydrogels. *ACS Macro Lett.* **2017**, *6*, 345–349. [CrossRef]
19. Uetani, K.; Koga, H.; Nogi, M. Checkered Films of Multiaxis Oriented Nanocelluloses by Liquid-Phase Three-Dimensional Patterning. *Nanomaterials* **2020**, *10*, 958. [CrossRef] [PubMed]
20. Uetani, K.; Izakura, S.; Koga, H.; Nogi, M. Thermal Diffusivity Modulation Driven by the Interfacial Elastic Dynamics Between Cellulose Nanofibers. *Nanoscale Adv.* **2020**, *2*, 1024–1030. [CrossRef]
21. Fan, Y.; Saito, T.; Isogai, A. Individual chitin nano-whiskers prepared from partially deacetylated α-chitin by fibril surface cationization. *Carbohydr. Polym.* **2010**, *79*, 1046–1051. [CrossRef]
22. Uetani, K.; Hatori, K. Thermal conductivity analysis and applications of nanocellulose materials. *Sci. Technol. Adv. Mater.* **2017**, *18*, 877–892. [CrossRef] [PubMed]
23. Tolaimate, A.; Desbrieres, J.; Rhazia, M.; Alagui, A. Contribution to the preparation of chitins and chitosans with controlled physico-chemical properties. *Polymer* **2003**, *44*, 7939–7952. [CrossRef]
24. Zhang, Y.; Zhang, X.; Ding, R.; Zhang, J.; Liu, J. Determination of the degree of deacetylation of chitosan by potentiometric titration preceded by enzymatic pretreatment. *Carbohydr. Polym.* **2011**, *83*, 813–817. [CrossRef]
25. Ifuku, S.; Nogi, M.; Abe, K.; Yoshioka, M.; Morimoto, M.; Saimoto, H.; Yano, H. Preparation of Chitin Nanofibers with a Uniform Width as α-Chitin from Crab Shells. *Biomacromolecules* **2009**, *10*, 1584–1588. [CrossRef]
26. Fan, Y.; Saito, T.; Isogai, A. Chitin Nanocrystals Prepared by TEMPO-Mediated Oxidation of α-Chitin. *Biomacromolecules* **2008**, *9*, 192–198. [CrossRef] [PubMed]
27. Xu, J.; Liu, L.; Yu, J.; Zou, Y.; Wang, Z.; Fan, Y. DDA (degree of deacetylation) and pH-dependent antibacterial properties of chitin nanofibers against Escherichia coli. *Cellulose* **2019**, *26*, 2279–2290. [CrossRef]

 *nanomaterials*

Article

Molecular and Crystal Structure of a Chitosan—Zinc Chloride Complex

Toshifumi Yui [1,*] , **Takuya Uto** [2] and **Kozo Ogawa** [3]

[1]   Faculty of Engineering, University of Miyazaki, Nishi 1-1 Gakuen-kibanadai, Miyazaki 889-2192, Japan
[2]   Organization for Promotion of Tenure Track, University of Miyazaki, Nishi 1-1 Gakuen-kibanadai, Miyazaki 889-2192, Japan; t.uto@cc.miyazaki-u.ac.jp
[3]   Research Institute for Advanced Science and Technology, Osaka Prefecture University, 1-2 Gakuencho, Sakai, Osaka 599-8570, Japan; ogawakt@kawachi.zaq.ne.jp
*   Correspondence: tyui@cc.miyazaki-u.ac.jp; Tel.: +81-985-58-7319

Abstract: We determined the molecular and packing structure of a chitosan–$ZnCl_2$ complex by X-ray diffraction and linked-atom least-squares. Eight D-glucosamine residues—composed of four chitosan chains with two-fold helical symmetry, and four $ZnCl_2$ molecules—were packed in a rectangular unit cell with dimensions $a = 1.1677$ nm, $b = 1.7991$ nm, and $c = 1.0307$ nm (where c is the fiber axis). We performed exhaustive structure searches by examining all of the possible chain packing modes. We also comprehensively searched the positions and spatial orientations of the $ZnCl_2$ molecules. Chitosan chains of antiparallel polarity formed zigzag-shaped chain sheets, where N2$\cdots$O6, N2$\cdots$N2, and O6$\cdots$O6 intermolecular hydrogen bonds connected the neighboring chains. We further refined the packing positions of the $ZnCl_2$ molecules by theoretical calculations of the crystal models, which suggested a possible coordination scheme of Zn(II) with an O6 atom.

Keywords: chitosan—$ZnCl_2$ complex; crystal structure; X-ray fiber diffraction

Citation: Yui, T.; Uto, T.; Ogawa, K. Molecular and Crystal Structure of a Chitosan—Zinc Chloride Complex. *Nanomaterials* **2021**, *11*, 1407. https://doi.org/10.3390/nano11061407

Academic Editors: Josefina Pons and Jorge Pasán

Received: 10 May 2021
Accepted: 24 May 2021
Published: 26 May 2021

Publisher's Note: MDPI stays neutral with regard to jurisdictional claims in published maps and institutional affiliations.

1. Introduction

Chitin—a linear polysaccharide composed of β-(1→4)-linked *N*-acetyl-D-glucosamine monomers—is the most abundant and renewable natural polymer after cellulose. One commonly finds chitin in the exoskeleton or cuticles of many invertebrates, and the cell walls of most fungi and some algae. Recently, chitinous skeletal fibers from some marine demosponges were attracted attention as a possible application to scaffolds for cultivation [1]. Chitosan, a partially or fully deacetylated derivative of chitin, exhibits a regular distribution of aliphatic primary amino groups and primary hydroxyl groups. Unlike its corresponding parent polymer, chitosan is soluble in various aqueous acids and has a remarkable ability to form specific complexes with a number of ions, such as transition and post-transition metal ions [2,3]. Chitosan is a unique cationic biopolymer that is available in large quantities. Researchers have commercialized it as follows: a coagulant for wastewater, animal feed, fertilizer, an antibacterial additive for clothing, and a precursor for glucosamine [4]. Chitosan is conventionally a powder or a film, yet chitosan nanofibers—typically fabricated by electrospinning—are an active area of research. Researchers have used electrospun nanofibers of chitosan—and corresponding blends with other polymers—for separations, as biological scaffolds, and for drug delivery [5–7]. Researchers have also used electrospun chitosan nanofibers to remove low concentrations of metal ions [8] and arsenate ions [9] from water.

Researchers first obtained the X-ray fiber pattern of chitosan—by solid-state deacetylation of a lobster tendon—in a hydrated crystalline form [10]. The results indicated a two-fold helical symmetry for chitosan as a molecular chain structure in a similar manner as chitin. Researchers first obtained the other polymorph of chitosan, an anhydrous form, by annealing the following: a stretched chitosan sample at ≥190 °C in water [11], and a crab tendon chitosan in water at ~240 °C [12]. The reported crystal structures of both the hydrated and anhydrous forms are in atomistic detail, as per analyses of corresponding X-ray diffraction data [12–14]. Researchers reported further structural details of the anhydrous form—such as the positions of the hydroxyl and amino hydrogen atoms—by using periodic density functional theory (DFT), which indicated a possible scheme of the hydrogen bond network [15]. The structures of chitosan metal complexes and salts are also pertinent. Researchers have proposed two types of coordination mode: the pendant model [16,17] and the bridge model [18]. The chitosan–HI salt, the only chitosan complex for which there is a crystal structure in atomistic detail, involves two independent iodide ions: one coordinated with three amino groups, and one that accepts one hydrogen atom from an amino group and two hydrogen atoms from the O6 hydroxyl groups [19]. This may be interpreted as a hybrid of the bridge and pendant coordination modes, in terms of interactions with the amino groups. However, the coordination structures of chitosan–metal complexes remained unsolved.

Ogawa et al. reported an X-ray diffraction study of crab tendon chitosan samples complexed with Cd(II), Zn(II), and Cu(II) salts; the $ZnCl_2$ complex sample provided a fiber diffraction pattern of higher quality compared with the other chitosan metal salts [20]. The affinity of chitosan for Zn cations can be pertinent to applications. For example, regarding wastewater treatment, chitosan exhibits a medium affinity for Zn(II) compared with other divalent metallic ions—approximately one-third that for Cu(II) and almost equivalent to that for Cd(II), which enabled a chitosan film to remove zinc ions up to ~50% of the initial concentration in the effluent [21]. The chitosan–Zn complex exhibited broad-spectrum antimicrobial activities, especially against *Escherichia coli* [22]. Researchers incorporated a zinc electrodeposit—applied on a steel construction surface for protection and as a barrier against corrosion—with chitosan to form chitosan–zinc composite electrodeposits with enhanced antibacterial properties [23]. Nanoscale fabrication of chitosan further enhances these functionalities of zinc complexes.

In the present study, we determined and analyzed the crystal structure of a chitosan–$ZnCl_2$ complex to reveal the coordination structure of the zinc ion by X-ray fiber diffraction, which we combined with molecular mechanics (MM) and quantum mechanical modeling. The final crystal structure that the molecular chains arranged to form zigzag-shaped chain sheets along the *a*-axis, where the neighboring chains were connected by intermolecular hydrogen bonds involving the N2 and O6 atoms. $ZnCl_2$ molecules were located at the bending positions of the chain sheets. Although no clear coordinate bond with an amino group was detected, O6–H···Zn(II) coordinate bonds were suggested in the semi-empirical quantum mechanics (SEQM)-optimized crystal model.

2. Materials and Methods

2.1. Sample Preparation and X-ray Diffraction

Our method of obtaining X-ray fiber diffraction data was described in our previous study [16,20]. Briefly, tendon chitosan was prepared from a highly oriented chitin specimen of a crab tendon, *Chionecetes opilio* O. Fabricius, by *N*-deacetylation with 67% aqueous sodium hydroxide at 100 °C for 2 h under a nitrogen atmosphere. The degree of *N*-acetylation of the tendon chitosan was found to be 0% by measurement of a colloidal titration, and the viscosity average polymerization was 10,800 [16,20]. The tendon chitosan was soaked in aqueous $ZnCl_2$. The X-ray fiber diffraction patterns were recorded with a box camera equipped with an imaging plate (Fujifilm HR-III), at 76% relative humidity in a helium atmosphere, with a Rigaku Geigerflex diffractometer equipped with Ni-filtered Cu Kα radiation. The X-ray diffraction image was read with an imaging plate detector (Rigaku R-AXIS), and the three-dimensional intensity profile was analyzed with Surfer (Golden Software, Inc., Golden, CO, USA) for resolution of overlapped profiles, background removal, and calculations of peak intensities. A set of the measured intensities was corrected for the Lorentz and polarization factors to provide a set of observed structure factors, F_o.

The density of the tendon chitosan–ZnCl$_2$ complex was measured by flotation with a carbon tetrachloride–*m*-xylene solution.

2.2. Crystal Structure Analysis

Figure 1 shows the principal parameters to describe chitosan chain conformation and position in the unit cell, together with atom labeling. A molecular chain structure with two-fold helical symmetry was the observed form of the chitosan crystal structures and was not substantially changed throughout structure refinement except in terms of the orientations of the hydroxymethyl groups, χ_{O6}, which were defined by the O5–C5–C6–O6 sequence. The conformation usually prefers three staggered positions, which was termed as either *gauche–trans* (*gt*), *gauche–gauche* (*gg*), or *trans–gauche* (*tg*); the χ_{O6} values of the respective positions are 60°, −60°, and 180°. The positions of the ZnCl$_2$ molecules were associated with the primary amino groups of the glucosamine residues and were defined by a set of internal coordinates: the N2–Zn distance, d_{N2-Zn}; the C2–N2–Zn angle, θ_{Zn}; the rotation of the C1–N2 bond, χ_{Zn}; the N1–Zn–Cl1 angle, θ_{Cl1}; and the rotation of N2–Zn, χ_{Cl1}. The torsion angles, χ_{Zn} and χ_{Cl1}, were defined by the atom sequences of C1–C2–N2–Zn and C2–N2–Zn–Cl1, respectively. The initial position of the other chlorine atom, Cl2, was defined by reporting 180° as the Cl1–Zn–Cl2 angle, such that the ZnCl$_2$ molecule of an initially linear structure could bend in either direction during structure refinement.

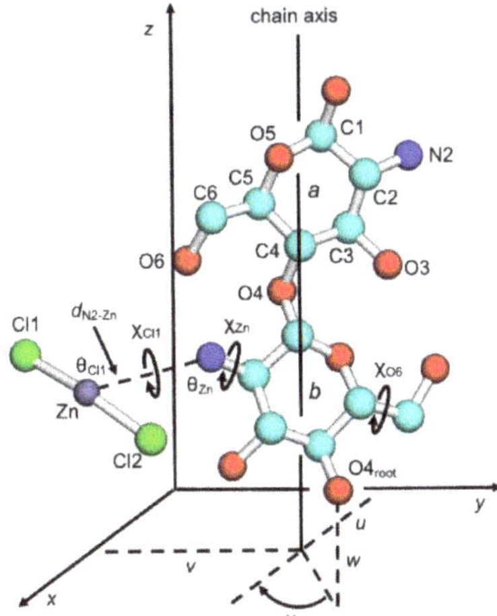

Figure 1. Two-fold helix structure of chitosan together with atom and residue designations. Notation is as follows: chain packing parameters (μ, u, v, and w), hydroxymethyl conformation (χ_{O6}), and ZnCl$_2$ position parameters (χ_{Zn}, χ_{Cl1}, θ_{Zn}, θ_{Cl1}, and d_{N2-Zn}).

The chain positions in the crystal unit cell were defined by the chain packing parameters: the chain rotational position, μ in degrees, with respect to the helix axis; and the chain translational positions, u, v, and w, in fractions, along the a, b, and c dimensions, respectively. Whereas the u and v parameters define the position of the helix axis on the ab base plane, the μ and w parameters correspond to the positions of the root atom, O4$_{root}$, in Figure 1; $\mu = 0°$ and $w = 0$ when O4$_{root}$ is at (x, 0, 0) for a helix origin.

Molecular and packing structures—of chitosan chains and ZnCl$_2$ molecules—were determined by using the linked-atom least-squares (LALS) program [24,25], where the quantity Ω was minimized as per the following:

$$\Omega = \sum_{m} w_m \left(|F_{m,o}|^2 - k^2 |F_{m,c}|^2 \right) + s \sum_{i,j} \epsilon_{i,j} + \sum_{q} \lambda_q G_q \tag{1}$$

The first summation term ensures optimum agreement between the observed ($F_{m,o}$) and calculated ($F_{m,c}$) X-ray structure amplitudes of the m-th reflection. Term k is a scaling factor. The weight of the reflection, w_m, was fixed to 1.0 for observed reflections, 0.5 for unobserved reflections in which $F_{m,c} > F_{m,o}$, and 0.0 for unobserved reflections in which $F_{m,c} < F_{m,o}$. The second summation term evaluates non-bonded repulsions between atoms i and j. The quantity s is an overall weight of the non-bonded repulsions. The third summation term imposes the atomic coordinate constraints by the method of Lagrange undermined multipliers. The constraints were adopted to preserve helix continuity and pyranose ring closure of the residue. The overall agreement between the observed and calculated X-ray structure amplitudes was evaluated by a weighed residual:

$$R_w = \frac{\sum_m w_m \left(|F_{m,o}| - |F_{m,c}| \right)^2}{\sum_m w_m F_o^2} \tag{2}$$

The unobserved reflections below the observable threshold were included in calculating both Equations (1) and (2). One-half of the minimum intensity was assigned to estimate the magnitude of F_o for unobserved reflections.

2.3. Theoretical Calculations of Crystal Models

A theoretical study of crystal models with finite dimensions was carried out to refine orientations of the hydroxyl, hydroxymethyl, and amino groups and the positions of the $ZnCl_2$ molecules. Two crystal models, differing in the constituent number of the chitosan chains, were constructed on the basis of the crystal structure determined by X-ray analysis: chitohexaose × 25 and chitohexaose × 9. Both models considered four $ZnCl_2$ molecules in the inner core, which enabled the $ZnCl_2$ molecules to exist in a crystalline state. Partial structure optimizations were applied to the crystal models, such that the three substituent group orientations and the $ZnCl_2$ positions were allowed to vary, whereas the backbone structures of chitohexaose were static. MM was applied to optimizations of the chitohexaose × 25 models by using a universal force field (UFF) [26] where the atomic charges were assigned by charge equilibration [27]. Researchers have applied a UFF to MM calculations involving metal complexes [27–29]. Optimizations of the chitohexaose × 9 models were performed by using SEQM with a PM6 Hamiltonian [30]. Tight convergence self-consistent field criteria (10^{-8} a.u.) were applied in the SEQM optimizations. The accuracy of the PM6 Hamiltonian for Zn(II) complexes was demonstrated by systematic benchmarks [30,31]. Its reliability was also reported for crystal structure studies of proteins and other organic materials with complexes of Zn(II) [32–34]. All of the MM and SEQM calculations were performed by using Gaussian 09 Rev. C01 [35].

2.4. Visualizations of Crystal Structures

Molecular graphics software, PyMOL 1.7.1, was used to visualize and draw the crystal structures [36].

3. Results and Discussion

3.1. Crystal Data

Figure 2 shows an X-ray fiber diffraction pattern of the chitosan–$ZnCl_2$ complex. We indexed a total of 33 observed diffraction spots up to the fourth layer with a rectangular unit cell, dimensions $a = 1.1677$ nm, $b = 1.7991$ nm, and $c = 1.0307$ nm (where c is the fiber axis). Table S1 shows the observed and calculated d-spacings. The unit cell accommodated eight glucosamine units—comprised of four chitosan chains with a two-fold helical conformation, and four $ZnCl_2$ molecules—resulting in a calculated density $\rho_{calc} = 1.41$ g/cm^3. One can insert an additional 16 water molecules into the unit cell, resulting in $\rho_{calc} = 1.63$ g/cm^3, for a better fit to the observed density, $\rho_{obs} = 1.56$ g/cm^3.

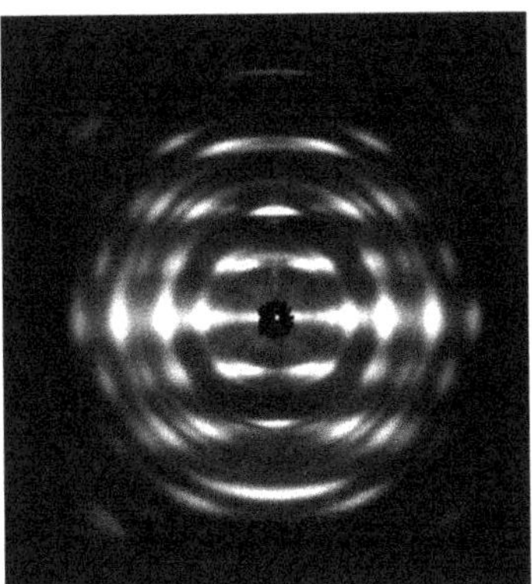

Figure 2. X-ray fiber diffraction pattern of the chitosan–ZnCl$_2$ complex.

3.2. Search for Chain Packing Structures

Researchers have proposed that chitosan chains are packed with $P2_12_12_1$ symmetry in the parent crystal structures of both the anhydrous chitosan and chitosan hydrate crystal structures, where the positioning of a pair of neighboring chains in antiparallel polarity is related by two-fold helical symmetry along either the *a* or *b* axis [12–14]. Similarly, we assumed involvement of two-fold helical symmetry to generate antiparallel chain pairs, despite the fact that we observed odd reflections on *h*00 and 0*k*0, such as 100 and 070, in the fiber diagram. We rationalized such violation of systematic absences by non-symmetrical packing of ZnCl$_2$ molecules. We assumed two types of space groups, $P2_1/a$ and $P2_1/b$, to define a chain packing structure in the unit cell. The space group models required two independent chains, each consisting of two consecutive glucosamine units as an asymmetric unit. We defined the eight chain packing models by a combination of the space group, chain positions, and chain polarities for the two independent chains (Table 1). For each of the chain packing models, we considered the three staggered positions of the hydroxymethyl group (*gt*, *gg*, and *tg*) to give 24 initial structures. For all eight glucosamine residues, we linked a ZnCl$_2$ molecule to an amino group with an initial $d_{\text{N2–Zn}}$ of 0.22 nm, whereas we set the occupancy of Zn and Cl atoms to be 0.5, and thus the number of ZnCl$_2$ molecules was effectively four, instead of eight.

Table 1. Chain packing modes of chain packing models.

Model	Chain Positions		Chain Polarities [1]	
	Chain 1 ($u1$, $v1$)	Chain 2 ($u2$, $v2$)	Chain 1	Chain 2
	$P2_1/a$ space group			
1	0.0, 0.0	0.0, 0.5	up	up
2			up	down
3	0.25, 0.25	0.25, 0.75	up	up
4			up	down
	$P2_1/b$ space group			
5	0.0, 0.0	0.5, 0.0	up	up
6			up	down
7	0.25, 0.25	0.25, 0.75	up	up
8			up	down

[1] When the position of C1 is higher than that of C4 along the fiber axis, a chain polarity is up. Otherwise, a chain polarity is down.

The first stage of the chain packing structure search was to determine the appropriate *ab* projection structures based on the $F(hk0)$ data for the initial 24 structures. The values of $\mu 1$ and $\mu 2$ were stepped from a value of $0°$ to a value of $180°$ by $10°$ increments to provide two-dimensional R_w maps. We generally found the R_w minima at approximately $(\mu 1, \mu 2)$ = $(0°, 0°)$, $(0°, 90°)$, $(90°, 0°)$, and $(90°, 90°)$ in most of the 24 R_w maps. On the basis of the results of the $\mu 1$–$\mu 2$ search, we generated 432 initial structures by combining six $\mu 1$ and $\mu 2$ values ($-10°$, $0°$, $10°$, $80°$, $90°$, and $100°$), three χ_{O6} values ($60°$, $-60°$, and $180°$), two χ_{Zn} values ($-60°$ and $120°$), and two χ_{Cl1} values ($0°$ and $90°$); and refined the structures with respect to the $\mu 1$ and $\mu 2$ values. Among the 432 $\mu 1$–$\mu 2$ refined structures, we selected that which exhibited the lowest R_w value to represent each of the chain packing models (Table S2). As a result, chain packing model 2 clearly differed from the other models by its comparatively low R_w values. Table 2 shows the final R_w values of the three models of chain packing model 2, all of which were adopted to search for an appropriate N2–Zn distance, $d_{\text{N2–Zn}}$. Because LALS sets bond length as a non-variable parameter, we carried out the $d_{\text{N2–Zn}}$ search by stepping the $d_{\text{N2–Zn}}$ value from a value of 0.22 nm to a value of 5.0 nm in 0.01-nm increments, whereas we varied the parameters $\mu 1$, $\mu 2$, χ_{O6}, χ_{Zn}, χ_{Cl1}, and θ_{Zn}. Table 2 shows the R_w values at the minima with respect to the $d_{\text{N2–Zn}}$ change, and the corresponding $d_{\text{N2–Zn}}$ values.

Table 2. Summary of *ab* projection structure analysis.

Hydroxymethyl Conformation	R_w ($d_{\text{N2–Zn}}$ in nm)			
	$\mu 1$-$\mu 2$ refinement			
gg	0.310 (0.22)			
gt	0.289 (0.22)			
tg	0.290 (0.22)			
	$d_{\text{N2–Zn}}$ search			
gg	0.192 (0.23)	0.392 (0.36)	0.302 (0.45)	
gt	0.185 (0.25)	0.150 (0.30)	0.199 (0.37)	0.208 (0.45)
tg	0.214 (0.23)	0.196 (0.40)	0.202 (0.46)	

We then focused our crystal structure analysis on three-dimensional chain packing structure searches using the higher-layers $F(hkl)$ data, where $l = 1 - 3$, in addition to the $F(hk0)$ data. For each of the 10 refined models (Table 2), we stepped the chain translational positions ($w1$ and $w2$) from a value of -0.4 to a value of 0.5 in increments of 0.1, which generated 100 structures with respect to the $w1$–$w2$ positions. We varied the $w1$ and $w2$ values, and the angle parameters θ_{Zn}, in addition to the parameters in the previous $d_{\text{N2–Zn}}$ search. We selected the three-dimensional chain packing structure with the lowest R_w value from the 100 refined structures. Among the remaining 10 models, we screened three ($d_{\text{N2–Zn}}$ = 0.30, 0.37, or 0.46 nm) for the next crystal structure refinement stage. Table 3 shows the final R_w values and the refined chain packing parameters of the three selected models. Although the initial hydroxymethyl conformations were *gt* for models 1 and 2, and *tg* for model 3 (Table 3), most of the χ_{O6} angles substantially rotated into different conformers compared with the original conformations.

Table 3. Chain packing parameters and R_w values of models selected in three-dimensional structure search.

Model	R_w	$d_{\text{N2–Zn}}$	Chain Packing Parameters			
			$\mu 1$ (deg.)	$w1$ (frac.)	$\mu 2$ (deg.)	$w2$ (frac.)
1	0.259	0.30	-11.7	0.068	116	-0.108
2	0.258	0.37	-2.91	-0.053	109	-0.174
3	0.245	0.46	-4.78	-0.214	109	-0.138

3.3. Crystal Structure Refinement by Combined X-ray Data and Stereochemical Constraints

We performed the final stage of crystal structure analysis mainly to refine the $ZnCl_2$ packing positions. We introduced the stereochemical constraints, the second term of Equation (1), to suppress development of unrealistically short contacts between nonbonding atoms during structure refinement. We excluded a hydroxymethyl conformation, χ_{O6}, from two-fold helical symmetry operation of the chain, which allowed four χ_{O6} parameters—corresponding to the two independent chains—to rotate independently. In the structure refinement stage, we set the occupancy of a $ZnCl_2$ molecule to be 1.0, such that the two amino groups were linked to the two respective $ZnCl_2$ molecules and the remaining two remained unlinked, which required six linking patterns of $ZnCl_2$ molecules (Figure S1). We allowed two sets of the $ZnCl_2$ position parameters (χ_{Zn}, χ_{Cl1}, and θ_{Zn}) to independently vary, and we stepped the remaining $ZnCl_2$ position parameter (d_{N2-Zn}) from a value of 0.27 nm to a value of 0.47 nm in 0.01 nm increments. At each d_{N-Zn} position, we generated 1296 initial structures by combining four sets of three χ_{O6} values ($60°$, $-60°$, and $180°$), two sets of two χ_{Zn} values ($-60°$ and $120°$), and two sets of two χ_{Cl1} values ($0°$ and $90°$); whereas we used the initial chain packing positions ($\mu1$, $\mu2$, $w1$, and $w2$) from either of the three selected models (Table 3), depending on the d_{N2-Zn} value. We carried out full minimization of the Ω function of Equation (1) with respect to all of the parameters previously discussed. Figure 3 shows the adiabatic R_w changes for the six $ZnCl_2$ linking pattern models with respect to the d_{N2-Zn} value, where we used the R_w values from the refined structures corresponding to the lowest R_w values among the 1296 possible values. The R_w values decreased in accordance with increasing d_{N-Zn}, accompanied by several minima, suggesting that the $ZnCl_2$ molecule preferentially intercalated between the chitosan chains rather than closely coordinated to a particular primary amino group. Table 4 shows the chain packing parameters (μ and w), R_w, and contact σ, the second summation term of Equation (1) representing a magnitude of short nonbonding contact, for the final seven structures, which we selected on the basis of the threshold R_w value of 0.245 (defined arbitrarily). Considering the fact that there were negligible differences in the R_w values among the seven final structures, we selected model 4 because it corresponded to the smallest contact σ value corresponding to a comparatively low R_w value.

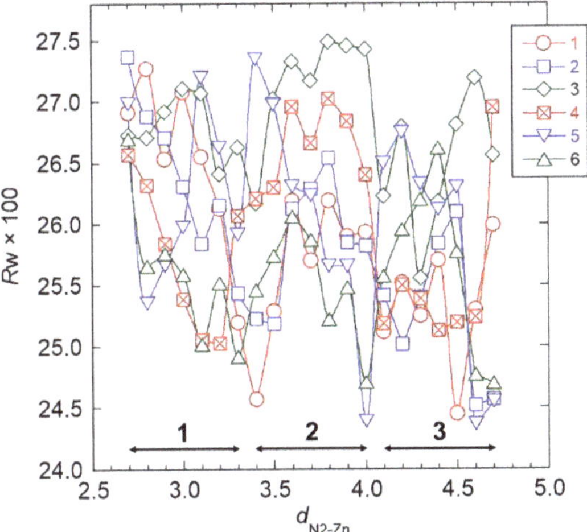

Figure 3. R_w with the respect to N2–Zn distance, d_{N2-Zn}, in structure refinement for six $ZnCl_2$ linking patterns. Double-headed arrows indicate d_{N2-Zn} ranges where we obtained the models by the preceding three-dimensional structure refinement. Numbers 1–6 refer to corresponding model numbers.

Table 4. Chain packing parameters, R_w, and nonbonding contacts of final refined models.

Model	ZnCl$_2$ Linking Pattern	Contact σ	R_w	d_{N2-Zn} (nm)	Chain Packing Parameters			
					$\mu 1$ (deg.)	$w1$ (frac.)	$\mu 2$ (deg.)	$w2$ (frac.)
1	1	83	0.246	0.34	3.61	−0.119	102	−0.243
2	5	138	0.244	0.40	−4.07	−0.116	106	−0.215
3	6	147	0.247	0.40	−0.402	−0.063	102	−0.209
4	1	52	0.245	0.45	12.1	−0.388	103	−0.345
5	2	172	0.245	0.46	11.5	−0.230	102	−0.169
6	5	193	0.244	0.46	−2.84	−0.214	107	−0.168
7	6	151	0.247	0.47	−2.02	−0.215	111	−0.154

We performed an additional full optimization run for model 4, where we introduced the angle parameter, θ_{Cl1}, to allow ZnCl$_2$ molecules to form a bent structure; we increased the magnitude of an overall weight of the non-bonded repulsions, s, in Equation (1) by 5× that which we adopted in the preceding runs to remove excessive non-bonding repulsions. Table 5 shows the values of the representative parameters of the final structure, and Figure 4 shows the structure's projections. Table S3 shows the coordinates of the final structure, and Table S4 shows the observed and calculated structure factor amplitudes.

Table 5. Representative parameters of the final structure.

Hydroxymethyl Conformations, χ_{O6}		
Labels	**Chain No.**	**Values (deg.)**
χ_{O6A}	1, 2	171
χ_{O6B}	1, 2	170
χ_{O6A}	3, 4	171
χ_{O6B}	3, 4	84.8

Chain Packing Parameters, μ and w		
Labels	**Chain no.**	**Values**
$\mu 1$/deg.	1, 3	14.01
$w1$/frac.	1, 3	−0.3853
$\mu 2$/deg.	2, 4	102.1
$w2$/frac.	2, 4	−0.3431

Distances of Intramolecular Hydrogen Bond		
Atom Labels		**Values (nm)**
O3a/b	O5a/b	0.272

Distances of Intermolecular Hydrogen Bonds				
Atom Labels	**Chain No.**	**Atom Labels**	**Chain No.**	**Values (nm)**
O6a	1	O6b	2	0.198
O6b	1	O6a	2	0.220
O6a	3	N2a	4	0.285
N2b	3	N2a	4	0.256
O6b	3	N2b	4	0.286
N2a	3	O6a	4	0.285
N2a	3	N2b	4	0.256
R_w				0.247
contact σ				39

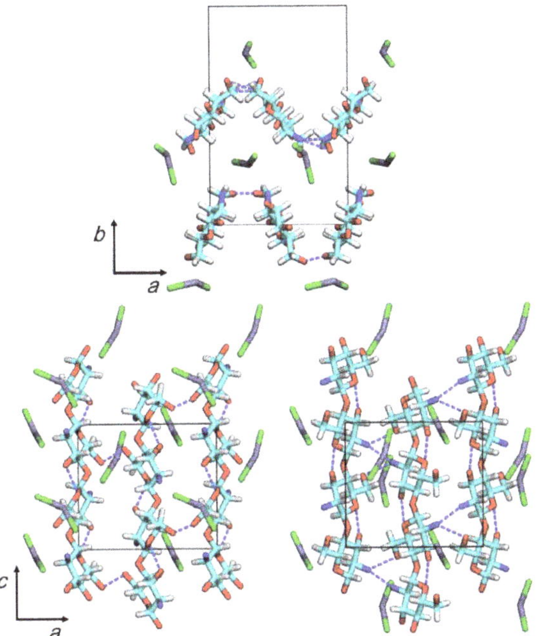

Figure 4. Projections of the proposed crystal structure on the *ab* (top) and *ac* (bottom) base planes (left: the chain sheet consisting of chains 1 and 2; right: the chain sheet consisting of chains 3 and 4). Dashed lines indicate hydrogen bonds.

Although the final R_W value increased slightly at the expense of a decreasing contact σ value, the short contact between an N2a atom in chain 3 and a Cl atom of the $ZnCl_2$–3 molecule remained, resulting in a distance of 0.22 nm. The N2 and O6 atoms participated in intermolecular hydrogen bonding to form chain sheets that exhibited a zigzag shape along the *a*-axis. The distances of O6a···O6b intermolecular hydrogen bonds, ~0.2 nm, involved the chain 1–2 sheets, possibly indicating the presence of short contacts in the atom pairs rather than hydrogen bonds. The other intermolecular hydrogen bonds in the chain 3–4 sheets formed with an optimum distance of ~0.28 nm. $ZnCl_2$ molecules were located around the bending portions of the chain sheets. The molecules were simply intercalated between the chain sheets and did not clearly coordinate to any particular amino or hydroxyl group nearby.

3.4. Theoretical Calculations of Crystal Models

We examined the crystal structure by using theoretical calculations implemented by MM and in accordance with SEQM methods, to refine the positions of the $ZnCl_2$ molecules as a primary objective. We performed the first MM calculations to search for preferred orientations of the hydroxyl and amino groups. Combining three staggered positions of the two hydroxyl groups on the C2 and C6 atoms, and those of the amino group, for each of the eight independent glucosamine units generated 6561 initial structures of the 25×hexaose model. We initially set the C3 hydroxyl group to be C2–C3–O3–H at *trans*, which formed an O3–H···O5 intramolecular hydrogen bond commonly observed in the parent chitosan crystal structures [12,13]. We partially optimized the structures of the 25×hexaose models, where the backbone structures of the chitosan chains were static, whereas we allowed the following to change: the orientations of the hydroxyl, amino, and hydroxymethyl groups; and the positions of the $ZnCl_2$ molecules. We selected 100 structures in terms of the lowest total steric energy and transferred their structural features (the substituent orientations and $ZnCl_2$ positions) to the 9 × hexaose models, which we then subjected to SEQM partial structure optimization. Figure 5 shows the *ab* projection of the lowest structure obtained by SEQM calculations. Compared with the starting crystal structure, the molecular axes of the $ZnCl_2$ molecules were more aligned with the fiber axis, whereas the Zn atoms had not moved appreciably from their initial positions. They had slightly bent to give Cl1–Zn–Cl2 angles of 147° and 152°, respectively, and the corresponding Zn atoms were coordinated by the adjacent O6–H groups, with N2–O6 distances of 0.22 and 0.24 nm, respectively. The DFT study of the D-glucosamine–$ZnCl_2$ complex models predicted the length of the coordinate bond between a glucosamine residue and a $ZnCl_2$ molecule to be ~0.20 nm [37]. The hydrogen-bonding network suggested for the crystal structure was found mostly conserved in the SEQM-optimized structure, where the amino groups served as hydrogen donors to the O6–H groups. Unfortunately, we observed all of the 100 SEQM-optimized structures—as per structure amplitudes—to give unacceptable R_w values of ~0.5. A possible interpretation for the considerable disagreement between the X-ray data and SEQM-optimized structure is that $ZnCl_2$ molecules, in the presence of water molecules, may partly dissociate in the chitosan–$ZnCl_2$ crystal structure. The crystal structures derived from concentrated $ZnCl_2$ solutions depend on complete coordination of Cl^- and H_2O around Zn^{2+} [38]. In the crystal structure of $ZnCl_2·2.5H_2O$, for example, one Zn^{2+} involved a tetrahedral coordination with Cl^-, and the other Zn^{2+} resided in the octahedral environment defined by five H_2O molecules and one Cl^- shared with $[ZnCl_4]^{2-}$ [38].

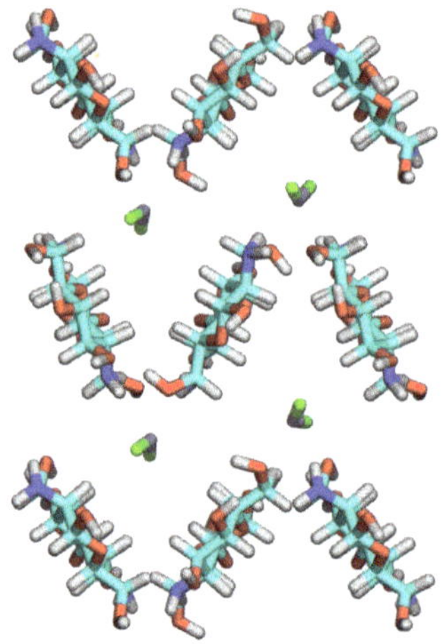

Figure 5. Projections of the SEQM-optimized structure of the crystal model on the *ab* base planes.

4. Conclusions

We reported atomistic detail of a chitosan–$ZnCl_2$ crystal structure. The molecular chains arrange to form zigzag-shaped chain sheets along the *a*-axis, where the neighboring chains in antiparallel polarity—related by two-fold helical symmetry along the same axis—are connected by intermolecular hydrogen bonds involving the N2 and O6 atoms. $ZnCl_2$ molecules are located at the bending positions of the chain sheets. The features are substantially similar to those detected in the crystal structure of chitosan–HI salt [19], suggesting that the features are likely common among chitosan–metal and chitosan–salt complexes. Although we did not detect a clear coordinate bond in the present crystal structure, a minor adjustment of the hydroxymethyl substituent structure may correspond to O6–H···Zn(II) coordinate bonds, in accordance with the structural features of the SEQM-optimized crystal model. It is possible that there is more stable complexation in the crystal structure involving dissociated Zn^{2+} and Cl^- ions, as well as H_2O molecules, resulting in Zn(II) that is readily accessible to the amino and hydroxyl groups. Two-fold helical symmetry—introduced to the present structure analysis because of technical reasons—restricted the diversity of the molecular packing, likely preventing glucosamine residues and ZnCl$_2$ molecules from complex formation. Extension to a complete $P1$ symmetry analysis may require one to investigate a large number of structural parameters and X-ray diffraction data obtained at higher resolution.

Supplementary Materials: The following are available online at https://www.mdpi.com/article/10.3390/nano11061407/s1, Figure S1: Six ZnCl$_2$ linking patterns for two independent chains; Table S1: Calculated and observed *d*-spacings; Table S2: R_w values of refined models in the chain packing search; Table S3: Fractional atomic coordinates of the final crystal structure; Table S4: Observed and calculated structure factor amplitudes of the final structure.

Author Contributions: Conceptualization, T.Y. and K.O.; sample preparation and X-ray diffraction measurement, K.O.; crystal structure analysis, T.Y.; theoretical calculations, T.U.; writing—review and editing, T.Y. All authors have read and agreed to the published version of the manuscript.

Funding: This research received no external funding.

Acknowledgments: The SEQM calculations were performed in part by using the Research Center for Computational Science (RCCS), Okazaki Research Facilities, and National Institutes of Natural Sciences, Japan. We thank Michael Scott Long, from Edanz (https://www.jp.edanz.com/ac (accessed on 25 May 2021)) for editing a draft of this manuscript.

Conflicts of Interest: The authors declare no conflict of interest.

References

1. Binnewerg, B.; Schubert, M.; Voronkina, A.; Muzychka, L.; Wysokowski, M.; Petrenko, I.; Djurovic, M.; Kovalchuk, V.; Tsurkan, M.; Martinovic, R.; et al. Marine biomaterials: Biomimetic and pharmacological potential of cultivated *Aplysina aerophoba* marine demosponge. *Mater. Sci. Eng. C* **2020**, *109*, 110566–110577. [CrossRef] [PubMed]
2. Muzzarelli, R.A.A. *Chitin*; Pergamon Press: Oxford, UK, 1977.
3. Kurita, K.; Sannan, T.; Iwakura, Y. Studies on chitin. VI. Binding of metal cations. *J. Appl. Polym. Sci.* **1979**, *23*, 511–515. [CrossRef]
4. Takeshita, S.; Zhao, S.; Malfait, W.J.; Koebel, M.M. Chemistry of chitosan aerogels: Three-dimensional pore control for tailored applications. *Angew. Chem. Int. Ed.* **2020**, *60*, 9828–9851. [CrossRef] [PubMed]
5. Ohkawa, K.; Minato, K.; Kumagai, G.; Hayashi, S.; Yamamoto, H. Chitosan nanofiber. *Biomacromolecules* **2006**, *7*, 3291–3294. [CrossRef] [PubMed]
6. Elsabee, M.Z.; Naguib, H.F.; Morsi, R.E. Chitosan based nanofibers, review. *Mater. Sci. Eng. C* **2012**, *32*, 1711–1726. [CrossRef]
7. AL-Jbour, N.D.; Beg, M.D.; Gimbun, J.; Alam, A. An overview of chitosan nanofibers and their applications in the drug delivery process. *Curr. Drug Deliv.* **2019**, *16*, 272–294. [CrossRef]
8. Horzum, N.; Boyaci, E.; Eroglu, A.E.; Shahwan, T.; Demir, M.M. Sorption efficiency of chitosan nanofibers toward metal ions at low concentrations. *Biomacromolecules* **2010**, *11*, 3301–3308. [CrossRef]
9. Min, L.L.; Zhong, L.B.; Zheng, Y.M.; Liu, Q.; Yuan, Z.H.; Yang, L.M. Functionalized chitosan electrospun nanofiber for effective removal of trace arsenate from water. *Sci. Rep.* **2016**, *6*, 32480. [CrossRef]
10. Clark, G.L.; Parker, E.A. An X-ray diffraction study of the action of liquid ammonia on cellulose and its derivatives. *J. Phys. Chem.* **1937**, *41*, 777–786. [CrossRef]
11. Ogawa, K.; Hirano, S.; Miyanishi, T.; Yui, T.; Watanabe, T. A new polymorph of chitosan. *Macromolecules* **1984**, *17*, 973–975. [CrossRef]
12. Yui, T.; Imada, K.; Okuyama, K.; Obata, Y.; Suzuki, K.; Ogawa, K. Molecular and crystal structure of the anhydrous form of chitosan. *Macromolecules* **1994**, *27*, 7601–7605. [CrossRef]
13. Okuyama, K.; Noguchi, K.; Miyazawa, T.; Yui, T.; Ogawa, K. Molecular and crystal structure of hydrated chitosan. *Macromolecules* **1997**, *30*, 5849–5855. [CrossRef]
14. Okuyama, K.; Noguchi, K.; Hanafusa, Y.; Osawa, K.; Ogawa, K. Structural study of anhydrous tendon chitosan obtained via chitosan/acetic acid complex. *Int. J. Biol. Macromol.* **1999**, *26*, 285–293. [CrossRef]
15. Ogawa, Y.; Kimura, S.; Wada, M.; Kuga, S. Crystal analysis and high-resolution imaging of microfibrillar α-chitin from *Phaeocystis*. *J. Struct. Biol.* **2010**, *171*, 111–116. [CrossRef]
16. Ogawa, K.; Oka, K.; Miyanishi, T.; Hirano, S. X-ray Diffraction Study on Chitosan-Metal Complexes. In *Chitin, Chitosan, and Related Enzymes*; Zikakis, J.P., Ed.; Academic Press: London, UK, 1984; pp. 327–345.
17. Domard, A. pH and c.d. measurements on a fully deacetylated chitosan: Application to Cu(II)—Polymer interactions. *Int. J. Biol. Macromol.* **1987**, *9*, 98–104. [CrossRef]

18. Schlick, S. Binding sites of Cu^{2+} in chitin and chitosan. An electron spin resonance study. *Macromolecules* **1986**, *19*, 192–195. [CrossRef]

19. Lertworasirikul, A.; Yokoyama, S.; Noguchi, K.; Ogawa, K.; Okuyama, K. Molecular and crystal structures of chitosan/HI type I salt determined by X-ray fiber diffraction. *Carbohydr. Res.* **2004**, *339*, 825–833. [CrossRef] [PubMed]

20. Ogawa, K.; Oka, K.; Yui, T. X-ray study of chitosan-transition metal complexes. *Chem. Mater.* **1993**, *5*, 726–728. [CrossRef]

21. Rhazi, M.; Desbrières, J.; Tolaimate, A.; Rinaudo, M.; Vottero, P.; Alagui, A.; El Meray, M. Influence of the nature of the metal ions on the complexation with chitosan. *Eur. Polym. J.* **2002**, *38*, 1523–1530. [CrossRef]

22. Wang, X.; Du, Y.; Liu, H. Preparation, characterization and antimicrobial activity of chitosan–Zn complex. *Carbohydr. Polym.* **2004**, *56*, 21–26. [CrossRef]

23. Zhai, X.; Sun, C.; Li, K.; Guan, F.; Liu, X.; Duan, J.; Hou, B. Synthesis and characterization of chitosan–zinc composite electrodeposits with enhanced antibacterial properties. *RSC Adv.* **2016**, *6*, 46081–46088. [CrossRef]

24. Arnott, S.; Scott, W.E. Accurate X-ray diffraction analysis of fibrous polysaccharides containing pyranose rings. Part I. The linked-atom approach. *J. Chem. Soc. Perkin Trans. 2* **1972**, *3*, 324–335. [CrossRef]

25. Smith, P.J.C.; Arnott, S. LALS: A linked-atom least-squares reciprocal-space refinement system incorporating stereochemical restraints to supplement sparse diffraction data. *Acta Crystallogr. Sect. A* **1978**, *34*, 3–11. [CrossRef]

26. Rappe, A.K.; Casewit, C.J.; Colwell, K.S.; Goddard, W.A.; Skiff, W.M. UFF, a full periodic table force field for molecular mechanics and molecular dynamics simulations. *J. Am. Chem. Soc.* **1992**, *114*, 10024–10035. [CrossRef]

27. Rappe, A.K.; Colwell, K.S.; Casewit, C.J. Application of a universal force field to metal complexes. *Inorg. Chem.* **1993**, *32*, 3438–3450. [CrossRef]

28. Deria, P.; Gomez-Gualdron, D.A.; Hod, I.; Snurr, R.Q.; Hupp, J.T.; Farha, O.K. Framework-topology-dependent catalytic activity of zirconium-based (porphinato)zinc(II) MOFs. *J. Am. Chem. Soc.* **2016**, *138*, 14449–14457. [CrossRef]

29. Zarabadi-Poor, P.; Marek, R. In silico study of (Mn, Fe, Co, Ni, Zn)-btc metal–organic frameworks for recovering xenon from exhaled anesthetic gas. *ACS Sustain. Chem. Eng.* **2018**, *6*, 15001–15006. [CrossRef]

30. Stewart, J.J. Optimization of parameters for semiempirical methods V: Modification of NDDO approximations and application to 70 elements. *J. Mol. Model.* **2007**, *13*, 1173–1213. [CrossRef]

31. Amin, E.A.; Truhlar, D.G. Zn coordination chemistry: Development of benchmark suites for geometries, dipole moments, and bond dissociation energies and their use to test and validate density functionals and molecular orbital theory. *J. Chem. Theory Comput.* **2008**, *4*, 75–85. [CrossRef]

32. Narayanan, V.V.; Gopalan, R.S.; Chakrabarty, D.; Mobin, S.M.; Mathur, P. Crystal structure and energy optimization of dichlorobis(ethylanthranilatonicotinamide)zinc(II). *J. Chem. Crystallogr.* **2011**, *41*, 801–805. [CrossRef]

33. Romero, M.J.; Suarez, V.; Fernandez-Farina, S.; Maneiro, M.; Martinez-Nunez, E.; Zaragoza, G.; Gonzalez-Noya, A.M.; Pedrido, R. Effect of the metal ion on the enantioselectivity and linkage isomerization of thiosemicarbazone helicates. *Chem. Eur. J.* **2017**, *23*, 4884–4892. [CrossRef] [PubMed]

34. Pinheiro, P.S.M.; Rodrigues, D.A.; Sant'Anna, C.M.R.; Fraga, C.A.M. Modeling zinc-oxygen coordination in histone deacetylase: A comparison of semiempirical methods performance. *Int. J. Quantum Chem.* **2018**, *118*, e25720. [CrossRef]

35. Frisch, M.J.; Trucks, G.W.; Schlegel, H.B.; Scuseria, G.E.; Robb, M.A.; Cheeseman, J.R.; Scalmani, G.; Barone, V.; Mennucci, B.; Petersson, G.A.; et al. *Gaussian 09, Revision C. 01*; Gaussian Inc.: Wallingford, UK, 2009.

36. Schrödinger LLC. *The PyMOL Molecular Graphics System, ver. 1.7.1*; Schrödinger LLC: New York, NY, USA, 2014.

37. Gomes, J.R.B.; Jorge, M.; Gomes, P. Interaction of chitosan and chitin with Ni, Cu and Zn ions: A computational study. *J. Chem. Thermodyn.* **2014**, *73*, 121–129. [CrossRef]

38. Hennings, E.; Schmidt, H.; Voigt, W. Crystal structures of $ZnCl_2 \cdot 2.5H_2O$, $ZnCl_2 \cdot 3H_2O$ and $ZnCl_2 \cdot 4.5H_2O$. *Acta Crystallogr. Sect. Sect. E Struct. Rep. Online* **2014**, *70*, 515–518. [CrossRef]

 *nanomaterials*

Article

Combination of Polysaccharide Nanofibers Derived from Cellulose and Chitin Promotes the Adhesion, Migration and Proliferation of Mouse Fibroblast Cells

Tomoka Noda, Mayumi Hatakeyama and Takuya Kitaoka *

Department of Agro-Environmental Sciences, Graduate School of Bioresource and Bioenvironmental Sciences, Kyushu University, Fukuoka 819-0395, Japan; t.noda@agr.kyushu-u.ac.jp (T.N.); m_hatakeyama@agr.kyushu-u.ac.jp (M.H.)
* Correspondence: tkitaoka@agr.kyushu-u.ac.jp; Tel.: +81-92-802-4665

Abstract: Extracellular matrix (ECM) as a structural and biochemical scaffold to surrounding cells plays significant roles in cell adhesion, migration, proliferation and differentiation. Herein, we show the novel combination of TEMPO-oxidized cellulose nanofiber (TOCNF) and surface-*N*-deacetylated chitin nanofiber (SDCtNF), respectively, having carboxylate and amine groups on each crystalline surface, for mouse fibroblast cell culture. The TOCNF/SDCtNF composite scaffolds demonstrated characteristic cellular behavior, strongly depending on the molar ratios of carboxylates and amines of polysaccharide NFs. Pure TOCNF substrate exhibited good cell attachment, although intact carboxylate-free CNF made no contribution to cell adhesion. By contrast, pure SDCtNF induced crucial cell aggregation to form spheroids; nevertheless, the combination of TOCNF and SDCtNF enhanced cell attachment and subsequent proliferation. Molecular blend of carboxymethylcellulose and acid-soluble chitosan made nearly no contribution to cell culture behavior. The wound healing assay revealed that the polysaccharide combination markedly promoted skin repair for wound healing. Both of TOCNF and SDCtNF possessed rigid nanofiber nanoarchitectures with native crystalline forms and regularly-repeated functional groups, of which such structural characteristics would provide a potential for developing cell culture scaffolds having ECM functions, possibly promoting good cellular adhesion, migration and growth in the designated cellular microenvironments.

Keywords: cellulose nanofiber; chitin nanofiber; surface carboxylation; surface deacetylation; cell culture scaffold; skin repair; wound healing; biomedical applications

Citation: Noda, T.; Hatakeyama, M.; Kitaoka, T. Combination of Polysaccharide Nanofibers Derived from Cellulose and Chitin Promotes the Adhesion, Migration and Proliferation of Mouse Fibroblast Cells. *Nanomaterials* **2022**, *12*, 402. https://doi.org/10.3390/nano12030402

Academic Editor: Wei Zhang

Received: 27 December 2021
Accepted: 21 January 2022
Published: 26 January 2022

Publisher's Note: MDPI stays neutral with regard to jurisdictional claims in published maps and institutional affiliations.

1. Introduction

An extracellular matrix (ECM) is a non-cellular biocomponent filled up in the intercellular spaces within tissues and organs, which is well known as structural and biochemical scaffolding constituents, such as collagen with a rigid nanofibrous protein, proteoglycans composed of core proteins and glycosaminoglycan chains, and a linear polysaccharide, e.g., hyaluronan [1]. These biological components found in vivo provide the structural frameworks for cell adhesion and growth, subsequently affecting proliferation and differentiation of the attached cells [2]. In recent years, the development of cell culture scaffolds that mimic the ECM components and cell-surrounding microenvironments has been actively carried out [3]. At the beginning, intact collagen from animal sources was used to promote cell attachment and angiogenesis in tissue engineering [4]. In a similar way, natural hyaluronan was also effective for wound healing when cultured with chondrocytes around damaged human cartilages [5]. However, such ECM components are often derived from animal origins, resulting in a risk of various infectious diseases and immune responses [6]. Furthermore, the extracted components are crude mixtures whose quality at chemistry level is not always constant; therefore, the lot-to-lot variation critically reduces reproducibility for expected performances. For that reason, novel xenobiotic but biocompatible components with various ECM functions have been required to design bioadaptive cell culture scaffolds, and in that case structurally-defined rigid nanofibers are promising candidates for this purpose.

Cellulose, the major structural component of wood cell walls, is the most abundant biopolymer on the earth, which forms crystalline rigid nanofibers composed of dozens of linear β-1,4-linked D-glucopyranose chains assembled by intra- and inter-molecular hydrogen bonding [7,8]. In recent years, cellulose nanofibers (CNFs) have attracted much attention in academic and industrial circles due to their high strength, light weight, high transparency, low thermal expansion and various fascinating features [9–11]. Most of CNFs can be isolated from wood and pulp fibers by physicochemical downsizing to a nanometer level. One approach to obtain CNFs is an aqueous catalytic oxidation of cellulose sources with 2,2,6,6-tetramethylpiperidine 1-oxyl (TEMPO) [12]. The TEMPO-mediated oxidation is allowed to site-selectively convert hydroxymethyl group to carboxylate one at the C6-position exposed on the crystalline surface of cellulose microfibrils [13]. The TEMPO-oxidized cellulose nanofiber (TOCNF) has a unique core–shell structure with intact cellulose in the core and polyuronate in the shell, and thus it is one of the most promising nanomaterials in eco-friendly and sustainable industries. Chitin (Ct) is also a natural biopolymer commonly found in shells of marine crustaceans and cell walls of fungi [14]. It is a typical liner polysaccharide, composed of *N*-acetyl D-glucosamine linked only in a β-1,4 manner like cellulose. Chitosan (Cs), a deacetylated derivative of Ct, is a positively-charged natural polymer that has unique bioactive properties such as antibacterial activity [15], wound healing [16] and analgesic effects [17]. Lately, chitin nanofiber (CtNF) [18], chitosan nanofiber (CsNF) [18] and surface *N*-deacetylated CtNF (SDCtNF) [19], which are obtained by nano-pulverization and stepwise deacetylation, are expected in the biomedical applications because of their structural properties as well as CNFs [20]. Especially, SDCtNF also has a unique core–shell structure with intact Ct in the core and Cs in the shell, similar to the nanoarchitecture of the TOCNF [19].

A variety of cell culture scaffolds composed of polysaccharides have been actively investigated [21–23]. An attempt on the molecular blend of Cs and hyaluronan was carried out to fabricate hybrid fibers as a scaffolding biomaterial for cartilage tissue engineering [24], on which chondrocytes well proliferated while maintaining their morphological phenotype and producing ECMs, especially type II collagen, around the cells. In addition, Cs and collagen composites have been investigated to promote osteoblast proliferation, differentiation and matrix mineralization for MC3T3-E1 cell culture [25]. Besides, such combination drastically increased transcriptional activity of Runx2, which is an important factor to regulate the downstream of osteoblast differentiation of phosphorylated Erk1/2 [25]. Our previous works have reported that glyco-biointerfaces composed both of chitohexaose (βGlcNAc6) and cellohexaose (βGlc6) promoted the activation of a specific detoxification enzyme of human hepatocellular carcinoma (HepG2) cells [26]. Furthermore, we unveiled that the inflammatory response of human embryonic kidney (HEK293) cells strongly depended on the surface βGlcNAc6 density, via direct stimulation triggered by toll-like receptor 2 [27]. However, neither molecular blends nor oligosaccharides assembly have any nanofiber structures, which are found in vivo around cells. TOCNF-containing hydrogels showed non-cytotoxicity for rat bone marrow-derived mesenchymal stem cells, and L929 fibroblast cells in vitro, and the hydrogels significantly reduced peritoneal adhesion in rats compared to untreated controls by in vivo evaluations [28]. Besides, in recent years, we have reported the carboxylate content-dependent cell proliferation of mouse fibroblasts by using TOCNF-based cell culture scaffolds, on which cellular behavior varied according to the surface physicochemistry [29]. Thus, rigid polysaccharide nanofibers must have the potential to develop a new type of bioadaptive cell culture scaffolds.

In this work, the novel combination of crystalline TOCNF and SDCtNF having carboxylate and primary amine groups, respectively, on each nanofiber surface was first investigated to develop bioadaptive cell culture scaffolds, as illustrated in Figure 1. We aimed at elucidating the characteristic cellular behavior induced by mixing polysaccharide NFs with different ratios and providing a potential to use these structurally-defined polysaccharide NFs as a new candidate for developing cell culture scaffolds. Rigid nanofibers and repeated functional groups of nano-polysaccharides demonstrated the unique cell attachment, migration and proliferation behavior of mouse fibroblast cells, although molecular blending of water-soluble polysaccharides analogous to the nanofibers used was ineffective. Our strategy is expected to provide a new insight into biomaterials design in cell culture engineering.

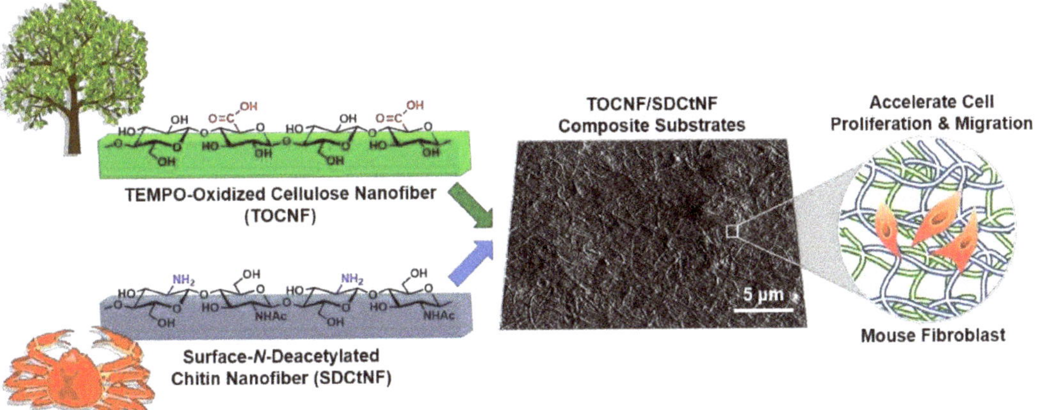

Figure 1. Schematic illustration of the research strategy in this work for the combination of TEMPO-oxidized cellulose nanofiber (TOCNF) and surface *N*-deacetylated chitin nanofiber (SDCtNF) to form crystalline-nanofibers-based cell culture scaffolds.

2. Experimental

2.1. Materials

TOCNF (0.93 wt% aqueous suspension, COOH: 1.59 mmol g^{-1}) and SDCtNF (1.0 wt% aqueous suspension, NH$_2$: 1.71 mmol g^{-1}) were kindly provided, respectively, by Nippon Paper Industries Co., Ltd., Tokyo, Japan and Marine Nano-fiber Co., Ltd., Tottori, Japan. Mouse fibroblast-like cell line (NIH/3T3) was purchased from RIKEN BRC, Tsukuba, Japan, through the National BioResource Project of the MEXT/AMED. Dulbecco's modified Eagle's medium (DMEM, high glucose), L-glutamine, penicillin–streptomycin, sodium pyruvate solution, and trypsin–ethylenediaminetetraacetic acid (EDTA) were obtained from Life Technologies Co., Carlsbad, CA, USA. Fetal bovine serum (FBS) was received from Biowest Co., Ltd., France. Micro-cover glass (diameter: 15 mm, Matsunami Glass Ind. Ltd., Osaka, Japan) was used as a base material to form polysaccharide cast films. Tissue culture polystyrene (TCPS) 24-well plates (Sumitomo Bakelite Co., Ltd., Tokyo, Japan) was used as cell culture substrate. Calcein AM (4 mM DMSO solution) for green fluorescent staining of live cells, and ethidium homodimer III (2 mM DMSO/H$_2$O 1:4 *v/v* solution) for fluorescent staining in red of dead cells were purchased from PromoCell GmbH, Germany (Live/Dead Cell Staining Kit II). Cell Counting Kit-8 (CCK-8; Dojindo Laboratories Ltd., Kumamoto, Japan) was used to measure the numbers of living cells. The water used in this study was purified with a Barnstead Smart2Pure system (Thermo Scientific Co., Ltd., Japan). Other reagents were of a special grade and used as received without further purification.

2.2. Preparation of Cell Culture Scaffolds from Polysaccharide Nanofibers

Each TOCNF or SDCtNF aqueous suspension was re-suspended to set at 0.4 wt% by adding purified water. The aqueous suspensions of nanofibers, TOCNF alone, SDCtNF alone, and the mixtures of TOCNF and SDCtNF with a molar ratio of COOH:NH$_2$ = 1:1,

2:1 and 4:1, were prepared. Each suspension was sufficiently dispersed to afford clear suspension with the use of an ultrasonic homogenizer (Ultra homogenizer VP-5S, TAITEC Co., Koshigaya, Japan) for 30 s, and casted on a micro-cover glass in an amount of 200 µL (nanofiber content: 0.8 mg in total), followed by being air-dried overnight at room temperature. Table S1 lists the actual amounts of each nanofiber coated on the micro-cover glass. The dried substrates were sterilized by immersion in ethanol with UV light for 20 min inside a clean bench.

2.3. Characterization of Polysaccharide Nanofibers and Substrates

The morphology of polysaccharide nanofibers used in this study, TOCNF, SDCtNF and TOCNF/SDCtNF, was observed by a transmission electron microscope (TEM; JEM-2100HC, JEOL Ltd., Tokyo, Japan) at an accelerating voltage of 200 kV. The diluted aqueous suspension was dropped onto a copper grid (elastic carbon coated, ELS-C10 STEM Cu100P grid specification, Ohken Shoji Co., Ltd., Tokyo, Japan), and then dyed with 1% sodium phosphotungstate for 3 min before the TEM observation. The surface nanotopography of the substrates was observed at room temperature in air using an atomic force microscope (AFM, Dimension Icon, Bruker Japan Co., Ltd., Tokyo, Japan) at a peak force tapping (ScanAsyst) mode. The measurement was performed with a scanning range of 20×20 µm^2 using a silicon nitride probe. Root mean square roughness (Ra) was calculated from obtained AFM images. The crystalline structures of freeze-dried samples were analyzed using an X-ray diffractometer (XRD; SmartLab, Rigaku Co., Ltd., Tokyo, Japan) with Cu Kα radiation (λ = 0.15418 nm). The XRD patterns were recorded at 40 kV within a scan range of 5° to 40° and a scan rate of 2° per min. The crystallinity indices were calculated in accordance with the Segal method [30]. Optical images of the coated films were taken by a digital camera. The surface wettability was analyzed by the sessile drop method using a contact angle meter (DropMaster 500, Kyowa Interface Science Co., Ltd., Niiza, Japan).

2.4. Cell Culture and Counting

Mouse fibroblast NIH/3T3 cells were cultured in DMEM supplemented with FBS (10%, v/v), penicillin–streptomycin (100 U and 100 µg mL^{-1}, respectively) at 37 °C in a humidified atmosphere of 5% CO_2 and 95% air. Each sterilized substrate was gently placed at the bottom of commercial 24-well TCPS plates with the top inner and bottom inner diameters of 16.3 mm and 15.1 mm, respectively. A total of 500 µL of cell suspension (1.0×10^5 cells mL^{-1}) was seeded on each substrate, and cultured at 37 °C. After incubation for 24, 48 and 72 h, the cultured cells were observed with a Leica DMI 4000B microscope (Leica Microsystems GmbH, Wetzlar, Germany). The number of living cells was measured by CCK-8 assay. Prior to cell counting, CCK-8 solution was added to each well by 25 µL, and incubated for 1.5 h (37 °C and 5% CO_2). After 1.5 h incubation, 200 µL solution was transferred from each well to a 96-well plate, and then absorbance at 450 nm was measured using a microplate reader (iMark microplate reader, Bio-Rad Laboratories Inc., Hercules, CA, USA). For the obtained absorbance, the number of living cells was quantified using a calibration curve made in advance.

2.5. Cell Assays

Cell viability on each substrate was confirmed by fluorescence observation after cell staining. After removing the medium from each well and washing with 1 mL of phosphate buffered saline (PBS), 200 µL of the staining solution containing 2 µM calcein AM and 4 µM ethidium homodimer III solutions in PBS, were added to each well, and then incubated for 30 min at 37 °C and 5% CO_2. Live cells were stained in green, and dead ones in red. Wound healing assay was carried out using Culture-Insert 2 Well (ibidi GmbH, Germany) according to the manufacturer's instruction. In brief, the Culture-Insert 2 Well having a 500-µm thickness was placed on the polysaccharide NFs-coated surface, while preventing any cell growth beneath the insert. NIH/3T3 cells were seeded on the separated two reservoirs in the well at a density of 2.8×10^4 cells per each reservoir. After allowing the cells to attach overnight, the culture-insert was removed, followed by adding a fresh medium. It was confirmed that the polysaccharide NFs layers remained intact without peeling off after removing the culture insert. Width of cell-free gap was set at 500 ± 100 μm, and cell migration behavior was monitored over time using a digital microscope (WSL-1800 CytoWatcher, ATTO Co., Ltd., Tokyo, Japan). Cell-free gap areas were captured by Image J software (National Institutes of Health, Bethesda, MD, USA) with Wound Healing Size Tool (an ImageJ/Fiji plugin) [31]. The percentage of wound closure was expressed as coverage percentage of wound closure on an initial area basis, according to the following equation:

$$\text{Wound closure (\%)} = \left[\frac{(At_0 - At)}{At_0} \right] \times 100$$

where At_0 is the initial cell-free area at time 0, and At is the cell-free area observed at 4, 8, 12, 16, 20 and 24 h.

3. Results and Discussion

3.1. Structural Characteristics of Polysaccharide Nanofibers and the Substrates

Nanoscale morphology of polysaccharide NFs used in this study was observed by TEM imaging. The obtained result clearly visualized independently-dispersed TOCNF and SDCtNF, having very thin fiber width, as shown in Figure 2a. Fiber width and ζ-potential values are listed in Table S1. Wood and crab-derived natural nanofibers are precisely constructed during biosynthesis, and nanomorphologies of TOCNF and SDCtNF corresponded well to the literature data [32]. Surface charges of TOCNF and SDCtNF at pH = 7.0 were ca. -46.8 mV and 17.5 mV, respectively, originating from dissociated caboxylate and protonated amine, while those of the mixtures depended on the molar ratios of $COOH:NH_2$. In the case of TOCNF/SDCtNF mixture, a lot of entangled nanofibers were observed, albeit being indistinguishable from each other. The XRD patterns of TOCNF, SDCtNF and the mixture are depicted in Figure 2b. TOCNF exhibited four major peaks at $2\theta = 14.8°, 16.4°, 22.6°$ and $34.2°$, corresponding to $(1–10)$, (110), (200) and (004) crystal planes of native cellulose I, respectively [33,34]. SDCtNF exhibited two typical peaks at $2\theta = 9°$ and $19°$, corresponding to (020) and (110) crystal planes of α-chitin [35,36]. The crystallinity index of TOCNF was ca. 66.6%, while that of SDCtNF was ca. 97.3%. Both strongly indicated the crystalline cores of each nanofiber, on which carboxylate $(1.59 \text{ mmol g}^{-1})$ and amine $(1.71 \text{ mmol g}^{-1})$ groups were present only on the surfaces for TOCNF [13] and SDCtNF [19], respectively. Therefore, TOCNF and SDCtNF used in this study presumably possessed unique core–shell structures composed of crystalline cores and functionalized shells. These nanofibers were actually very thin as compared to the natural ECM components such as collagen microfibers found in vivo, while they appeared similar to tropocollagen with 1.5-nm diameter and 300-nm length in nanomorphology [37]. In this context, TOCNF and SDCtNF were not directly involved in acting as structural analogues to native ECM components. On the other hand, the nanometer-scale topography is considered to affect the intra-/intercellular sensing functions [38,39]; therefore, thin TOCNF and SDCtNF would be expected to assume some influence on the cell behavior at the biointerfaces.

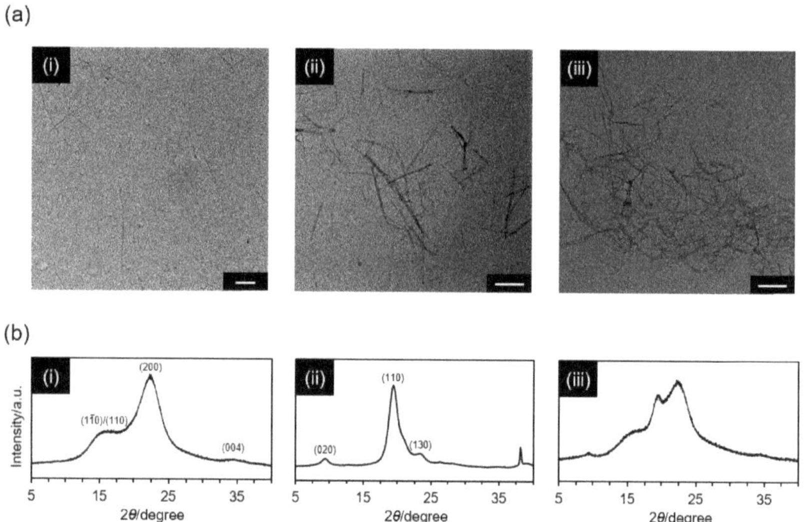

Figure 2. TEM images (**a**) and XRD patterns (**b**) of (**i**) TOCNF, (**ii**) SDCtNF and (**iii**) TOCNF/SDCtNF (COOH:NH$_2$ = 4:1). Scale bars in (**a**) = 200 nm.

Optical images and water wettability of polysaccharide NFs-coated glass substrates are shown in Figure 3a. Single-component substrates originating either from TOCNF or SDCtNF exhibited high transparency, whereas the TOCNF/SDCtNF composite substrate was a little bit translucent, possibly due to some aggregation. The surface wettability of cell culture scaffolds is a key issue to make an impact on cell attachment, and in general adequate hydrophobicity to promote the adsorption of adhesive proteins is required for practical cell culture [40]. The contact angles of a water droplet on each substrate, TOCNF alone, SDCtNF alone and TOCNF/SDCtNF composite, exhibited a hydrophilic surfaces, ranging from ca. 35° to ca. 50°, being regarded as a conventional bioinert surface, i.e., a non-cell-adhesive surface, as previously reported [41]. To determine the physical functionality of the substrates at a cell perception level, the surface roughness of the substrates was measured by AFM imaging. Figure 3b depicts the surface topography of each substrate, clearly indicating the accumulation of thin nanofibers to form dense network structures. The single-component TOCNF or SDCtNF substrates provided very flat surfaces with *Ra* = 2.94 nm and 4.06 nm, respectively, while the TOCNF/SDCtNF composite showed relatively rough surface with *Ra* = 53.4 nm, as shown in Table S1. This was attributed to the local aggregation induced by ionic crosslinking between negatively-charged carboxylates on the TOCNF and positively-charged amines on the SDCtNF [42,43], due to partial charge compensation presumed from the ζ-potential variation. Kunzler et al. reported that human gingival fibroblasts cultured on the same component substrate with different roughness at a micrometer level from 1 to 6 μm of *Ra* exhibited different morphological behavior of cells during proliferation [44]. The TOCNF/SDCtNF composites having ca. 50 nm of *Ra* were regarded as being very flat at the cell perception level. On the other hand, the nanometer-scale topography is considered to affect the intra-/intercellular sensing functions [38,39]. These polysaccharide NFs scaffolds possessed nanometer-scale roughness, and would possibly assume some influence on the cell behavior at the biointerfaces. Besides, the nanofibers-entangled structure was swollen but insoluble without adding any crosslinking agents in an aqueous medium during cell culture. NIH/3T3 fibroblast cells could proliferate on the surfaces of these nanofiber mats, which were much different from polymer films from a viewpoint of biointerface structures. However, the swollen structures of the nanofibers-entangled mats also may change the surface roughness in the culture medium, and therefore this concern will be investigated in our future work.

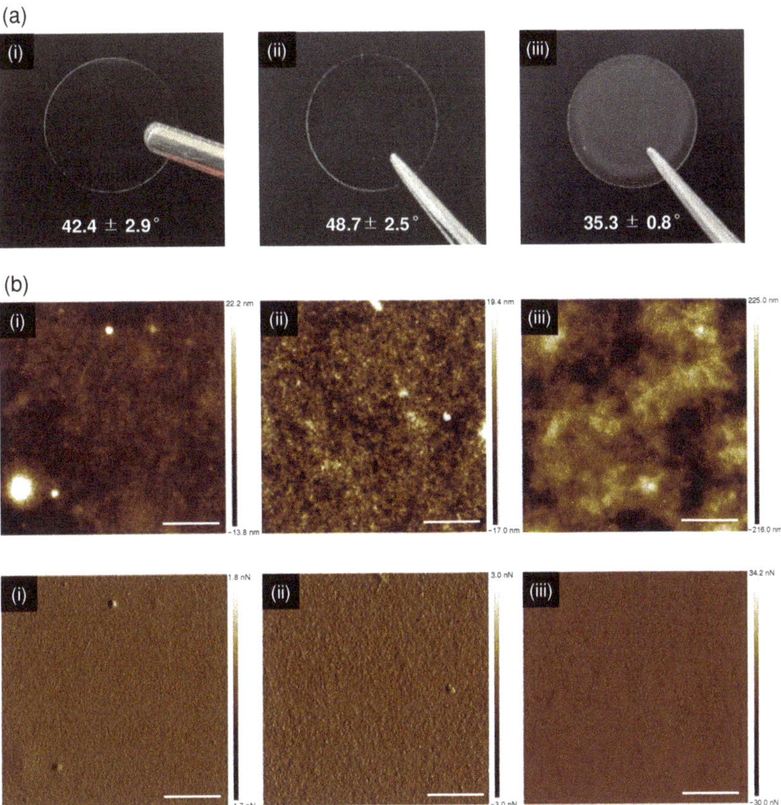

Figure 3. Optical images of polysaccharide NFs-coated glass substrates (**a**) and AFM images (**b**) of (**i**) TOCNF alone, (**ii**) SDCtNF alone and (**iii**) TOCNF/SDCtNF composite (COOH:NH$_2$ = 4:1). Numerical values in the optical images are the contact angles of a water droplet on each substrate after sterilization. **Upper** and **lower** images of (**b**) correspond to the topological and peak force error images, respectively. Scale bars in (**b**) = 5 μm.

3.2. Proliferation Behavior of Mouse Fibroblasts on Polysaccharide NFs Substrates

The effects of the combination of two polysaccharide NFs, TOCNF and SDCtNF, on cell viability and proliferation, were investigated using mouse fibroblast NIH/3T3 cells, which were subjected to cell culture either on single TOCNF, single SDCtNF or TOCNF/SDCtNF composite substrates. Figure 4a displays cell morphologies of NIH/3T3 cells with live/dead staining on each substrate after 72-h incubation. The NIH/3T3 cells adhered to and extended on the single TOCNF substrate as well as TCPS, reported in our previous study [29]; however, single SDCtNF substrate exhibited poor cell adhesion to form spheroid-like aggregates. The molar ratios of carboxylate and amine groups of the TOCNF/SDCtNF composites strongly influenced the cell growth behavior, as shown in Figure 4b and Figure S1. Increasing molar ratios of COOH:NH$_2$ from 1:1 to 4:1 in the TOCNF/SDCtNF composite substrates markedly improved the cellular attachment, from spheroid formation to cell spreading after 72-h culture. The TOCNF/SDCtNF composite substrate with COOH:NH$_2$ = 4:1 was superior to TCPS for cell proliferation, although polysaccharide NFs substrates exhibited hydrophilicity disadvantageous for conventional cell attachment. The surface charge modulates protein adsorption to direct integrin binding and specificity, thereby controlling cell adhesion. Thevenot et al. have reported that the incorporation of negative charges facilitated the adsorption of proteins which promoted cell adhesion and responses [45]. On the other hand, such strong interaction between material surfaces and cells made negative impacts on the cell growth rate due to strong cell adhesion. In this study, a single-component TOCNF substrate possesses a negative surface charge, ca. -46.8 mV of ζ-potential, possibly preferable for cell adhesion; however, it may interact strongly with the cultured cells. Thus, in the TOCNF/SDCtNF composite substrates, mixing TOCNF and SDCtNF having opposite charges, as shown in Table S1, was presumably allowed to tune the surface characteristics for cell adhesion and subsequent proliferation.

(a)   (b)

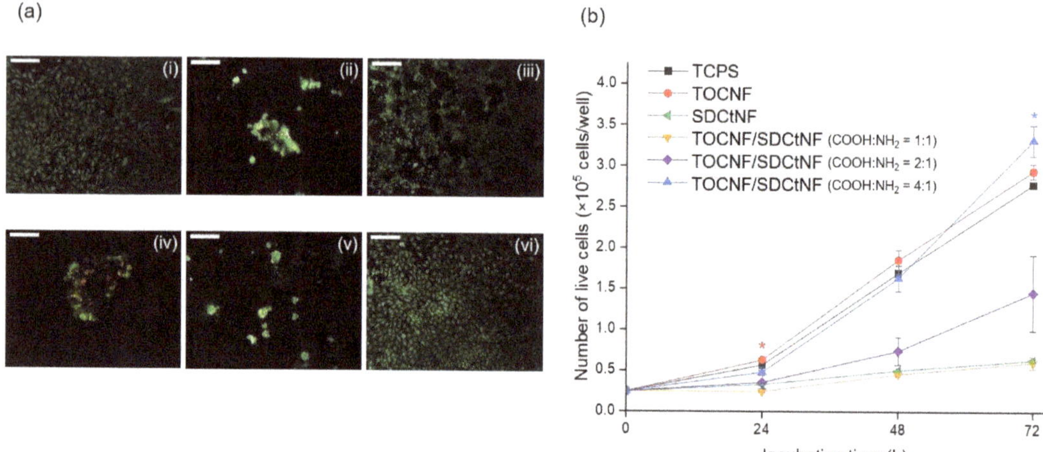

Figure 4. (**a**) Fluorescence images of NIH/3T3 cells cultured for 72 h of (**i**) TOCNF alone, (**ii**) SDCtNF alone, (**iii**) TCPS, (**iv**) TOCNF/SDCtNF (COOH:NH$_2$ = 1:1), (**v**) TOCNF/SDCtNF (2:1) and (**vi**) TOCNF/SDCtNF (4:1). Live cells were stained with calcein AM (green) and dead cells with ethidium homodimer III (red). Scale bars: 200 μm. (**b**) Cell numbers on each substrate after 24, 48 and 72 h of culture. Mean ± SD, $n = 3$, * $p < 0.05$ vs. TCPS.

The amine content of SDCtNF used here was 1.71 mmol g^{-1}, which was the maximum value as reported [19]; however, further N-deacetylation inside of CtNF was possible to form highly-deacetylated CsNF. Thus, another combination of TOCNF and commercial CsNF (NH$_2$: 4.35 mmol g^{-1}) was investigated as comparison. TEM image of CsNF exhibited thicker fibrous morphology than SDCtNF, and thin TOCNFs were entangled around the thick CsNF in the composite (Figure S2a). CsNF showed a typical XRD pattern of Cs; but the crystallinity was relatively low (Figure S2b). The surface roughness (Ra) of the CsNF substrate was much greater than SDCtNF and TOCNF (Figure S3). Therefore, the uniform substrates composed of TOCNF and CsNF could not be obtained by simple mixing. Nevertheless, molar ratios of carboxylate and amine of TOCNF/CsNF composites strongly affected the proliferation rate of NIH/3T3 cells (Figure S4). In this case, increasing molar ratios of COOH:NH$_2$ tended to decrease cell proliferation efficiency, where the COOH:NH$_2$ = 1:1 was the most preferable for cell growth (Figure S5). Direct comparison was a little difficult because the amine content of CsNF was 2.5 times greater than that of SDCtNF. The negatively-charged TOCNF possibly interacted with the outer surface of positively-charged NFs. The weight ratio of TOCNF and SDCtNF at COOH:NH$_2$ = 4:1 was almost the same as that of TOCNF and CsNF at COOH:NH$_2$ = 1:0.63, possibly assuming the similar situation for TOCNF-accessible amines exposed on the surfaces of SDCtNF and CsNF.

An important thing here was the fact that the simple molecular blend of CMC and Cs made no contribution to such unique cell proliferation behavior even at the molar ratios of COOH:NH$_2$ = 1:1 and 4:1, as shown in Figure S6. Although CMC and Cs had carboxylate and amine groups, respectively, similar as TOCNF and SDCtNF, the combination of CMC and Cs was not effective at all for cell culture, strongly indicating the significance of nanofiber forms in order to enhance the cell proliferation behavior. In vivo, the ECM biocomponents in general possess the nanofibrous forms, not molecular ones; therefore, a lot of research on cell culture substrates has paid attention to various synthetic and natural nanofibers [46,47]. The nanofibers-entangled structure was swollen but insoluble without any crosslinkers in an aqueous medium. NIH/3T3 fibroblast cells could proliferate on the surfaces of these nanofiber mats. It is not the purpose of this study to achieve complete imitation of natural ECM structures by using TOCNF and SDCtNF. Our polysaccharide NFs possessed solid-state interfaces, possibly interacting with the adhered cells at the nanometer level via stimulating intra-/intercellular sensing functions [38,39]. In this study, we have proposed to use TOCNF and SDCtNF, which both possess very fine, stable, natural-origin and xeno-free nanofiber architecture, and the combination of negatively-charged TOCNF and positively-charged SDCtNF is expected to tune the surface physicochemistry of the scaffold, directly affecting cell culture behavior.

3.3. Cell Migration Behavior for Wound Healing

Cell migration is an essential process for multicellular organisms, and indispensable for tissue developments, repair and regeneration [48]. Directional migration is in general triggered in response to extracellular stimuli such as chemokines and ECM components [49,50]. As found out above, the combination of TOCNF and SDCtNF could control the cell adhesion and proliferation behavior; therefore it was expected to manipulate cell migration. Thus, wound healing assay was investigated using mouse NIH/3T3 cells on our substrates. Figure 5a shows snapshot images of cell migration during wound closure of cell sheets with a gap of 500 μm fabricated on each substrate, and wound closure rates are shown in Figure 5b. Video S1 visualizes wound repair behavior for 48 h with animation of the obtained snapshots. TOCNF/SDCtNF substrate remarkably accelerated the wound closure process in the NIH/3T3 cell culture. Cell growth on the single-component SDCtNF substrate could not reach to be confluent, resulting in much difficulty in cell migration test (data not shown). The wound closure percentage on single TOCNF substrate varied from 13.5% for 12 h to 48.6% for 24 h, as compared to an initial area at t = 0, whereas control TCPS substrate exhibited the slow closure varying from 23.4% for 12 h to 60.6% for 24 h. Thus, it was presumably indicated that the NIH/3T3 cells strongly adhered to the surface of TOCNF rather than TCPS, resulting in less migration. On the other hand, the combination of TOCNF and SDCtNF with a molar ratio of $COOH:NH_2 = 4:1$ demonstrated rapid wound closure varying from 36.6% for 12 h to 78.1% for 24 h, presumably indicating the promotion of cell migration. Wound closure rate was faster for the combination of TOCNF and SDCtNF substrate than TOCNF alone and TCPS as control. Although single-component TOCNF substrate was effective for promoting the strong adhesion of NIH/3T3 fibroblasts to the scaffold surface, such interaction was not effective for cell migration, resulting in slow wound repair. Water-soluble Cs molecules have been investigated for fibroblast activation, cytokine production and stimulation of type IV collagen synthesis [51]; however poor cell adhesion causes problems for in vitro cell culture. Our strategy for the combination of natural polysaccharide nanofibers, TOCNF and SDCtNF, allowed to promote cell attachment, subsequent cell proliferation and smooth cell migration, which is expected to expand the possibility to tune the physicochemical and biological properties of xeno-free cell culture scaffolds.

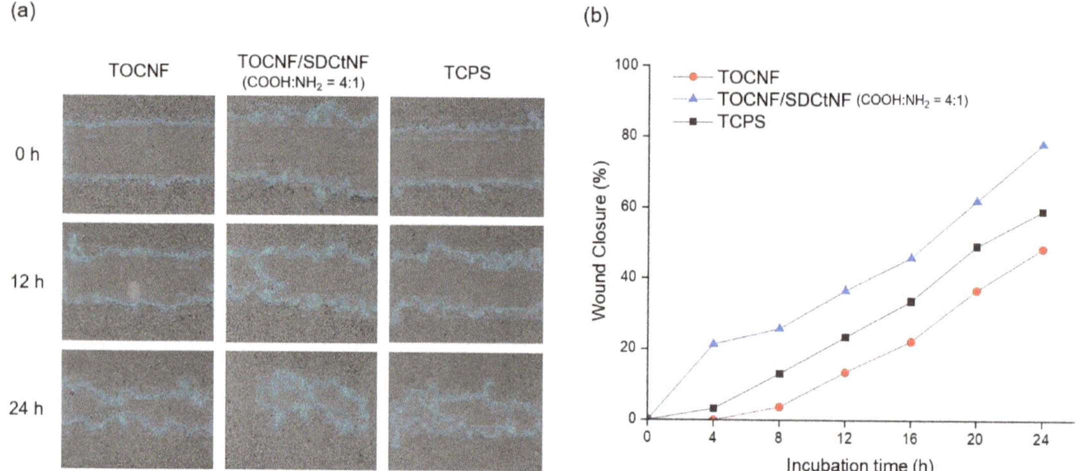

Figure 5. (**a**) Representative snapshot images of wound closure behavior of NIH/3T3 cells on different substrates. Cyan lines indicate the edges of the moving cells during the migration. (**b**) Wound closure rate of NIH/3T3 cells cultured on different substrates.

4. Conclusions

In conclusion, the combination of two types of polysaccharide nanofibers, TOCNF and SDCtNF, was effective for modulating cell attachment, subsequent proliferation and rapid cell migration due to the soft attachment of mouse fibroblast NIH/3T3 cells to the substrates. Fibroblast proliferation strongly depended on the molar ratios of functional groups, carboxylates and amines, respectively, present on TOCNF and SDCtNF surfaces. Molecular blend of water-soluble polysaccharides, analogous to TOCNF and SDCtNF, exhibited no positive effect on such unique cellular behavior. Thus, rigid nanofiber forms like ECM biocomponents found in vivo must be a key issue to manipulate the cellular response. Unique combination of structural natural polysaccharide nanofibers would be a promising strategy to design bioadaptive nanomaterials for tissue engineering.

Supplementary Materials: The following supporting information can be downloaded at: https://www.mdpi.com/article/10.3390/nano12030402/s1, Table S1: Characterization of polysaccharide NFs and substrates used in this study; Figure S1: Optical and fluorescence images of NIH/3T3 cells cultured on each substrate for 24, 48 and 72 h; Figure S2: TEM images of CsNF and TOCNF/CsNF (COOH:NH$_2$ = 1:1); XRD pattern of CsNF; Figure S3: Characterization of polysaccharide NFs-coated substrates; Optical and AFM images of CsNF alone and TOCNF/CsNF composite (COOH:NH$_2$ = 1:1); Contact angles of a water droplet on each substrate after sterilization; Figure S4: Optical and fluorescence images of NIH/3T3 cells cultured on TOCNF/CsNF substrate for 24, 48 and 72 h; Figure S5: Cell proliferation behavior on TOCNF/CsNF substrates; Time-course profiles of cell growth on TOCNF alone, CsNF alone and TOCNF/CsNF composite (COOH:NH$_2$ = 1:1); Effect of molar ratios of COOH and NH$_2$ in the TOCNF/CsNF composite substrates on cell proliferation at 72 h; Figure S6: Optical and fluorescence images of NIH/3T3 cells cultured on CMC/Cs substrates for 24, 48 and 72 h; Video S1: Wound closure behavior of NIH/3T3 cells on each substrate; TOCNF alone, TOCNF/SDCtNF composite (COOH:NH$_2$ = 4:1) and TCPS.

Author Contributions: T.N. conducted all experiments and analytical characterization. M.H. and T.K. conceived the conception of this work. T.K. designed research as a project administrator. T.N., M.H. and T.K. contributed to writing and reviewing the manuscript. All authors have read and agreed to the published version of the manuscript.

Funding: This research was funded by the Grant-in-Aid for Scientific Research (KAKENHI) Program (grant numbers JP20K22592 to M.H., JP21K14890 to M.H., JP21K19150 to T.K.) from the Japan Society for the Promotion of Science, and the Short-term Intensive Research Support Program from the Faculty of Agriculture, Kyushu University (M.H. and T.K.).

Institutional Review Board Statement: Not applicable.

Informed Consent Statement: Not applicable.

Data Availability Statement: Data presented in this study are available in this article.

Acknowledgments: The authors would like to thank Shinsuke Ifuku, Tottori University, who also belongs to Marine Nano-fiber Co., Ltd., Tottori, Japan, for providing free samples of SDCtNF. The authors appreciate technical assistance from the Center of Advanced Instrumental Analysis, Kyushu University; and the Ultramicroscopy Research Center, Kyushu University.

Conflicts of Interest: The authors declare no conflict of interest.

References

1. Frantz, C.; Stewart, K.M.; Weaver, V.M. The extracellular matrix at a glance. *J. Cell Sci.* **2010**, *123*, 4195–4200. [CrossRef]
2. Clause, K.C.; Barker, T.H. Extracellular matrix signaling in morphogenesis and repair. *Curr. Opin. Biotechnol.* **2013**, *24*, 830–833. [CrossRef] [PubMed]
3. Kim, T.G.; Shin, H.; Lim, D.W. Biomimetic scaffolds for tissue engineering. *Adv. Funct. Mater.* **2012**, *22*, 2446–2468. [CrossRef]
4. Chan, E.C.; Kuo, S.M.; Kong, A.M.; Morrison, W.A.; Dusting, G.J.; Mitchell, G.M.; Lim, S.Y.; Liu, G.S. Three dimensional collagen scaffold promotes intrinsic vascularisation for tissue engineering applications. *PLoS ONE* **2016**, *11*, e0149799. [CrossRef]
5. Albrecht, C.; Tichy, B.; Nürnberger, S.; Zak, L.; Handl, M.J.; Marlovits, S.; Aldrian, S. Influence of cryopreservation, cultivation time and patient's age on gene expression in Hyalograft ® C cartilage transplants. *Int. Orthop.* **2013**, *37*, 2297–2303. [CrossRef] [PubMed]
6. Razavi, M. *Biomaterials for Tissue Engineering*; Frontiers in Biomaterials; Bentham Science Publishers: Sharjah, United Arab Emirates, 2017; ISBN 9781681085371.
7. Kroon-Batenburg, L.M.J.; Kroon, J. The crystal and molecular structures of cellulose I and II. *Glycoconj. J.* **1997**, *14*, 677–690. [CrossRef] [PubMed]
8. Moon, R.J.; Martini, A.; Nairn, J.; Simonsen, J.; Youngblood, J. Cellulose nanomaterials review: Structure, properties and nanocomposites. *Chem. Soc. Rev.* **2011**, *40*, 3941–3994. [CrossRef]
9. Lavoine, N.; Bergström, L. Nanocellulose-based foams and aerogels: Processing, properties and applications. *J. Mater. Chem. A* **2017**, *5*, 16105–16117. [CrossRef]
10. Nascimento, D.M.; Nunes, Y.L.; Figueirêdo, M.C.B.; De Azeredo, H.M.C.; Aouada, F.A.; Feitosa, J.P.A.; Rosa, M.F.; Dufresne, A. Nanocellulose nanocomposite hydrogels: Technological and environmental issues. *Green Chem.* **2018**, *20*, 2428–2448. [CrossRef]
11. Sharma, A.; Thakur, M.; Bhattacharya, M.; Mandal, T.; Goswami, S. Commercial application of cellulose nano-composites—A review. *Biotechnol. Rep.* **2019**, *21*, e00316. [CrossRef]
12. Saito, T.; Nishiyama, Y.; Putaux, J.L.; Vignon, M.; Isogai, A. Homogeneous suspensions of individualized microfibrils from TEMPO-catalyzed oxidation of native cellulose. *Biomacromolecules* **2006**, *7*, 1687–1691. [CrossRef]
13. Saito, T.; Kimura, S.; Nishiyama, Y.; Isogai, A. Cellulose nanofibers prepared by TEMPO-mediated oxidation of native cellulose. *Biomacromolecules* **2007**, *8*, 2485–2491. [CrossRef] [PubMed]
14. Tokura, S.; Tamura, H. Chitin and Chitosan. *Compr. Glycosci. Chem. Syst. Biol.* **2007**, *2–4*, 449–475. [CrossRef]
15. Belbekhouche, S.; Bousserrhine, N.; Alphonse, V.; Le Floch, F.; Charif Mechiche, Y.; Menidjel, I.; Carbonnier, B. Chitosan based self-assembled nanocapsules as antibacterial agent. *Colloids Surf. B Biointerfaces* **2019**, *181*, 158–165. [CrossRef] [PubMed]
16. Dragostin, O.M.; Samal, S.K.; Dash, M.; Lupascu, F.; Pânzariu, A.; Tuchilus, C.; Ghetu, N.; Danciu, M.; Dubruel, P.; Pieptu, D.; et al. New antimicrobial chitosan derivatives for wound dressing applications. *Carbohydr. Polym.* **2016**, *141*, 28–40. [CrossRef]
17. Okamoto, Y.; Kawakami, K.; Miyatake, K.; Morimoto, M.; Shigemasa, Y.; Minami, S. Analgesic effects of chitin and chitosan. *Carbohydr. Polym.* **2002**, *49*, 249–252. [CrossRef]
18. Ifuku, S. Chitin and chitosan nanofibers: Preparation and chemical modifications. *Molecules* **2014**, *19*, 18367–18380. [CrossRef]
19. Fan, Y.; Saito, T.; Isogai, A. Individual chitin nano-whiskers prepared from partially deacetylated α-chitin by fibril surface cationization. *Carbohydr. Polym.* **2010**, *79*, 1046–1051. [CrossRef]
20. Azuma, K.; Ifuku, S.; Osaki, T.; Okamoto, Y.; Minami, S. Preparation and biomedical applications of chitin and chitosan nanofibers. *J. Biomed. Nanotechnol.* **2014**, *10*, 2891–2920. [CrossRef]
21. Zhai, P.; Peng, X.; Li, B.; Liu, Y.; Sun, H.; Li, X. The application of hyaluronic acid in bone regeneration. *Int. J. Biol. Macromol.* **2020**, *151*, 1224–1239. [CrossRef]
22. Aguilar, A.; Zein, N.; Harmouch, E.; Hafdi, B.; Bornert, F.; Offner, D.; Clauss, F.; Fioretti, F.; Huck, O.; Benkirane-Jessel, N.; et al. Application of Chitosan in Bone and Dental Engineering. *Molecules* **2019**, *24*, 3009. [CrossRef] [PubMed]

23. Dutta, S.D.; Patel, D.K.; Lim, K.-T. Functional cellulose-based hydrogels as extracellular matrices for tissue engineering. *J. Biol. Eng.* **2019**, *13*, 55. [CrossRef] [PubMed]
24. Yamane, S.; Iwasaki, N.; Majima, T.; Funakoshi, T.; Masuko, T.; Harada, K.; Minami, A.; Monde, K.; Nishimura, S. Feasibility of chitosan-based hyaluronic acid hybrid biomaterial for a novel scaffold in cartilage tissue engineering. *Biomaterials* **2005**, *26*, 611–619. [CrossRef] [PubMed]
25. Wang, X.; Wang, G.; Liu, L.; Zhang, D. The mechanism of a chitosan-collagen composite film used as biomaterial support for MC3T3-E1 cell differentiation. *Sci. Rep.* **2016**, *6*, 39322. [CrossRef] [PubMed]
26. Yoshiike, Y.; Kitaoka, T. Tailoring hybrid glyco-nanolayers composed of chitohexaose and cellohexaose for cell culture applications. *J. Mater. Chem.* **2011**, *21*, 11150–11158. [CrossRef]
27. Hatakeyama, M.; Nakada, F.; Ichinose, H.; Kitaoka, T. Direct stimulation of cellular immune response via TLR2 signaling triggered by contact with hybrid glyco-biointerfaces composed of chitohexaose and cellohexaose. *Colloids Surf. B Biointerfaces* **2019**, *175*, 517–522. [CrossRef]
28. Sultana, T.; Van Hai, H.; Abueva, C.; Kang, H.J.; Lee, S.Y.; Lee, B.T. TEMPO oxidized nano-cellulose containing thermo-responsive injectable hydrogel for post-surgical peritoneal tissue adhesion prevention. *Mater. Sci. Eng. C* **2019**, *102*, 12–21. [CrossRef]
29. Hatakeyama, M.; Kitaoka, T. Surface-Carboxylated Nanocellulose-Based Bioadaptive Scaffolds for Cell Culture. *Cellulose* **2021**, *5*, 1–15. [CrossRef]
30. Segal, L.; Creely, J.J.; Martin, A.E.; Conrad, C.M. An Empirical Method for Estimating the Degree of Crystallinity of Native Cellulose Using the X-Ray Diffractometer. *Text. Res. J.* **1959**, *29*, 786–794. [CrossRef]
31. Suarez-Arnedo, A.; Figueroa, F.T.; Clavijo, C.; Arbeláez, P.; Cruz, J.C.; Muñoz-Camargo, C. An image J plugin for the high throughput image analysis of in vitro scratch wound healing assays. *PLoS ONE* **2020**, *15*, e0232565. [CrossRef]
32. Isogai, A.; Saito, T.; Fukuzumi, H. TEMPO-oxidized cellulose nanofibers. *Nanoscale* **2011**, *3*, 71–85. [CrossRef] [PubMed]
33. Takahashi, Y.; Matsunaga, H. Crystal structure of native cellulose. *Macromolecules* **1991**, *24*, 3968–3969. [CrossRef]
34. Tang, Z.; Li, W.; Lin, X.; Xiao, H.; Miao, Q.; Huang, L.; Chen, L.; Wu, H. TEMPO-Oxidized cellulose with high degree of oxidation. *Polymers* **2017**, *9*, 421. [CrossRef] [PubMed]
35. Minke, R.; Blackwell, J. The structure of α-chitin. *J. Mol. Biol.* **1978**, *120*, 167–181. [CrossRef]
36. Ifuku, S.; Shervani, Z.; Saimoto, H. Chitin Nanofibers, Preparations and Applications. In *Adv. Nanofibers*; Maguire, R., Ed.; IntechOpen: London, UK, 2013; pp. 85–101. [CrossRef]
37. Stetefeld, J.; Frank, S.; Jenny, M.; Schulthess, T.; Kammerer, R.A.; Boudko, S.; Landwehr, R.; Okuyama, K.; Engel, J. Collagen Stabilization at Atomic Level. *Structure* **2003**, *11*, 339–346. [CrossRef]
38. Lord, M.S.; Foss, M.; Besenbacher, F. Influence of nanoscale surface topography on protein adsorption and cellular response. *Nano Today* **2010**, *5*, 66–78. [CrossRef]
39. Dalby, M.J.; Yarwood, S.J.; Riehle, M.O.; Johnstone, H.J.H.; Affrossman, S.; Curtis, A.S.G. Increasing Fibroblast Response to Materials Using Nanotopography: Morphological and Genetic Measurements of Cell Response to 13-nm-High Polymer Demixed Islands. *Exp. Cell Res.* **2002**, *276*, 1–9. [CrossRef]
40. Maroudas, N.G. Polymer exclusion, cell adhesion and membrane fusion. *Nature* **1975**, *254*, 695–696. [CrossRef]
41. Courtenay, J.C.; Johns, M.A.; Galembeck, F.; Deneke, C.; Lanzoni, E.M.; Costa, C.A.; Scott, J.L.; Sharma, R.I. Surface modified cellulose scaffolds for tissue engineering. *Cellulose* **2017**, *24*, 253–267. [CrossRef]
42. Grande, R.; Trovatti, E.; Carvalho, A.J.F.; Gandini, A. Continuous microfiber drawing by interfacial charge complexation between anionic cellulose nanofibers and cationic chitosan. *J. Mater. Chem. A* **2017**, *5*, 13098–13103. [CrossRef]
43. Pajorova, J.; Skogberg, A.; Hadraba, D.; Broz, A.; Travnickova, M.; Zikmundova, M.; Honkanen, M.; Hannula, M.; Lahtinen, P.; Tomkova, M.; et al. Cellulose Mesh with Charged Nanocellulose Coatings as a Promising Carrier of Skin and Stem Cells for Regenerative Applications. *Biomacromolecules* **2020**, *21*, 4857–4870. [CrossRef] [PubMed]
44. Kunzler, T.P.; Drobek, T.; Schuler, M.; Spencer, N.D. Systematic study of osteoblast and fibroblast response to roughness by means of surface-morphology gradients. *Biomaterials* **2007**, *28*, 2175–2182. [CrossRef]
45. Thevenot, P.; Hu, W.; Tang, L. Surface Chemistry Influences Implant Biocompatibility. *Curr. Top. Med. Chem.* **2008**, *8*, 270–280. [CrossRef]
46. Chen, S.; John, J.V.; McCarthy, A.; Xie, J. New forms of electrospun nanofiber materials for biomedical applications. *J. Mater. Chem. B* **2020**, *8*, 3733–3746. [CrossRef]
47. Fu, Q.; Duan, C.; Yan, Z.; Li, Y.; Si, Y.; Liu, L.; Yu, J.; Ding, B. Nanofiber-Based Hydrogels: Controllable Synthesis and Multifunctional Applications. *Macromol. Rapid Commun.* **2018**, *39*, 1800058. [CrossRef] [PubMed]
48. Trepat, X.; Chen, Z.; Jacobson, K. Cell migration. *Compr. Physiol.* **2012**, *2*, 2369–2392. [CrossRef] [PubMed]
49. Yue, B. Biology of the extracellular matrix: An overview. *J. Glaucoma* **2014**, *23*, S20–S23. [CrossRef]
50. Hughes, C.E.; Nibbs, R.J.B. A guide to chemokines and their receptors. *FEBS J.* **2018**, *285*, 2944–2971. [CrossRef]
51. Muzzarelli, R.A.A. Chitins and chitosans for the repair of wounded skin, nerve, cartilage and bone. *Carbohydr. Polym.* **2009**, *76*, 167–182. [CrossRef]

Article

Doxorubicin Embedded into Nanofibrillated Bacterial Cellulose (NFBC) Produces a Promising Therapeutic Outcome for Peritoneally Metastatic Gastric Cancer in Mice Models via Intraperitoneal Direct Injection

Hidenori Ando [1], Takashi Mochizuki [1], Amr S. Abu Lila [2,3], Shunsuke Akagi [1], Kenji Tajima [4], Kenji Fujita [1], Taro Shimizu [1], Yu Ishima [1], Tokuo Matsushima [5], Takatomo Kusano [5] and Tatsuhiro Ishida [1,*]

1   Department of Pharmacokinetics and Biopharmaceutics, Institute of Biomedical Sciences, Tokushima University, Tokushima 770-8505, Japan; h.ando@tokushima-u.ac.jp (H.A.); c401503025@tokushima-u.ac.jp (T.M.); c401603010@tokushima-u.ac.jp (S.A.); kenji-fujita@taiho.co.jp (K.F.); shimizu.tarou@tokushima-u.ac.jp (T.S.); ishima.yuu@tokushima-u.ac.jp (Y.I.)
2   Department of Pharmaceutics and Industrial Pharmacy, Faculty of Pharmacy, Zagazig University, Zagazig 44519, Egypt; amr_selim78@yahoo.com
3   Department of Pharmaceutics, College of Pharmacy, Hail University, Hail 81442, Saudi Arabia
4   Faculty of Engineering, Hokkaido University, Hokkaido 060-8628, Japan; ktajima@eng.hokudai.ac.jp
5   Kusano Sakko Inc., Hokkaido 067-0063, Japan; t-matsushima@kusanosk.co.jp (T.M.); tk@kusanosk.co.jp (T.K.)
*   Correspondence: ishida@tokushima-u.ac.jp; Tel.: +81-88-633-7260

Citation: Ando, H.; Mochizuki, T.; Lila, A.S.A.; Akagi, S.; Tajima, K.; Fujita, K.; Shimizu, T.; Ishima, Y.; Matsushima, T.; Kusano, T.; et al. Doxorubicin Embedded into Nanofibrillated Bacterial Cellulose (NFBC) Produces a Promising Therapeutic Outcome for Peritoneally Metastatic Gastric Cancer in Mice Models via Intraperitoneal Direct Injection. *Nanomaterials* **2021**, *11*, 1697. https://doi.org/10.3390/nano11071697

Academic Editor: Takuya Kitaoka

Received: 13 June 2021
Accepted: 25 June 2021
Published: 28 June 2021

Publisher's Note: MDPI stays neutral with regard to jurisdictional claims in published maps and institutional affiliations.

Abstract: Natural materials such as bacterial cellulose are gaining interest for their use as drug-delivery vehicles. Herein, the utility of nanofibrillated bacterial cellulose (NFBC), which is produced by culturing a cellulose-producing bacterium (*Gluconacetobacter intermedius* NEDO-01) in a medium supplemented with carboxymethylcellulose (CMC) that is referred to as CM-NFBC, is described. Recently, we demonstrated that intraperitoneal administration of paclitaxel (PTX)-containing CM-NFBC efficiently suppressed tumor growth in a peritoneally disseminated cancer xenograft model. In this study, to confirm the applicability of NFBC in cancer therapy, a chemotherapeutic agent, doxorubicin (DXR), embedded into CM-NFBC, was examined for its efficiency to treat a peritoneally disseminated gastric cancer via intraperitoneal administration. DXR was efficiently embedded into CM-NFBC (DXR/CM-NFBC). In an in vitro release experiment, 79.5% of DXR was released linearly into the peritoneal wash fluid over a period of 24 h. In the peritoneally disseminated gastric cancer xenograft model, intraperitoneal administration of DXR/CM-NFBC induced superior tumor growth inhibition (TGI = 85.5%) by day 35 post-tumor inoculation, compared to free DXR (TGI = 62.4%). In addition, compared with free DXR, the severe side effects that cause body weight loss were lessened via treatment with DXR/CM-NFBC. These results support the feasibility of CM-NFBC as a drug-delivery vehicle for various anticancer agents. This approach may lead to improved therapeutic outcomes for the treatment of intraperitoneally disseminated cancers.

Keywords: nanofibrillated bacterial cellulose; bacterial cellulose; peritoneally disseminated gastric cancer; doxorubicin; intraperitoneal chemotherapy

1. Introduction

The advent of nanotechnology has caused a profound impact on cancer therapeutics and diagnostics. Recently, there has been considerable progress in the development of nano-based drug delivery system that can specifically target the tumor site while minimizing damage to normal cells. Furthermore, a great shift in paradigm in regard to the use of nanobiomaterials over conventional systems has been observed. Currently, several types of biomaterials such as proteins, lipids, polysaccharides, polymers, etc., are widely used nowadays to deliver drugs to the tumor site [1].

Cellulose is a naturally produced homopolysaccharide composed of β-D-glucopyranose units linked by β-1,4-glycosidic bonds, and is mainly synthesized by plants [2,3]. Additionally, cellulose is also synthesized by strains of bacteria such as *Agrobacterium*, *Alcaligenes*, *Pseudomonas*, *Rhizobium*, *Sarcina*, and *Gluconacetobacter* [4–7]. Bacterial cellulose (BC), produced by such aerobic bacteria under static culture conditions, has recently received increased attention owing to unique physiochemical properties that are not found in plant cellulose. BC is more chemically pure and contains neither hemicellulose nor lignin. In addition, BC has nanometric characteristics, higher levels of water retention, higher levels of biocompatibility, and greater tensile strength that is the result of a greater amount of polymerization and an ultrafine network architecture [8–10].

By virtue of their unique properties, cellulose nanofibers (CNFs) such as BC have recently attracted a considerable amount of attention as drug delivery vehicles [11–13]. CNF aerogels, with favorable floatability and mucoadhesive properties, prepared by freeze-drying methods, were efficiently used for formulating gastroretentive drug delivery systems with improved bioavailability [11]. The potential of CNFs as delivery carriers for poorly soluble drugs has also been described [13]. Nevertheless, most studies have focused mainly on exploiting the unique physicochemical properties of CNFs that enhance drug loading in vitro, but little consideration has been given to the in vivo therapeutic efficacy of formulated drug-loaded CNFs.

In our recent study, paclitaxel (PTX)-embedded nanofibrillated bacterial cellulose (NFBC) were prepared for intraperitoneal administration and studied for their therapeutic efficacy on peritoneally disseminated gastric cancer in a xenograft nude mouse model. The results were compared with those from treatment with Taxol® (TAX) and Abraxane®, which are generally used in clinical settings. Systemic side effects such as those induced by the use of TAX were not present [14]. This represented the first report of a therapeutic effect of a BC-based anticancer drug formulation in a small animal model. The NFBC was synthesized using waste glycerol from a novel strain of *Gluconacetobacter intermedius* (named NEDO-01) isolated from the surfaces of fruits [15]. The NFBC showed potential for use as a drug-carrier vehicle. Under optimum *Gluconacetobacter* culture conditions, NEDO-01 produces a large amount of NFBC [15], which significantly reduces manufacturing cost and increases the possibility for practical application. By synthesizing NFBC in a medium supplemented with either carboxymethylcellulose (CMC) or hydroxypropylcellulose (HPC), the production of either hydrophilic NFBC (CM-NFBC) or amphiphilic NFBC (HP-NFBC) is possible [16]. In addition, in contrast to conventional top-down types of CNFs, which are obtained via mechanical treatment of wood pulp that contains submicro- or micro-sized fibers [17], NFBC synthesized via a bottom-up process from low-molecular-weight biomass via the use of NEDO-01 contains only trace amounts of submicro- or micro-sized fibers. Furthermore, manipulation of the culture conditions for the preparation of homogeneous nanofibers may be used to optimize the production of NFBC [15,16].

For this study, to extend the usability of NFBC as a drug-delivery vehicle, doxorubicin (DXR) was embedded in CM-NFBC (DXR/CM-NFBC). The in vivo therapeutic effects of the embedded material in a peritoneally disseminated gastric cancer xenograft mouse model were assessed.

2. Materials and Methods

2.1. Materials

DXR (Adryacin®) was purchased from Kyowa Kirin (Tokyo, Japan). CMC was kindly provided by Daicel (Osaka, Japan). *D*-luciferin potassium salt and bovine serum albumin (BSA) were purchased from FUJIFILM Wako Pure Chemical Corporation (Osaka, Japan). All other reagents were of analytical grade.

2.2. Animals

BALB/c mice (5-week-old, male) and BALB/c *nu/nu* mice (5-week-old, male) weighing 20–22 g were purchased from Japan SLC (Shizuoka, Japan). The small animals had free access to water and mouse chow and were housed under controlled environmental conditions (constant temperature, humidity and a 12 h dark/light cycle). All animal experiments were evaluated and approved by the Animal and Ethics Review Committee of Tokushima University (the approval code: T2019-47).

2.3. Preparation of CM-NFBC

CM-NFBC was prepared from an identified strain, NEDO-01, as previously described [16]. Briefly, the bacteria were inoculated into 10 mL Hestrin and Schramm's (HS) medium and incubated at 30 °C for 3 days under static conditions. The pellicle (3D network structure) formed at the air/liquid interface was pressed using a sterilized toothpick to isolate the bacteria. The culture medium containing bacteria (1 mL) was then inoculated into 10 mL of HS medium and incubated at 30 °C for 3 days under static conditions. The cultured medium containing bacteria (5 mL) was further inoculated into 100 mL HS medium supplemented with 2% w/v CMC and incubated at 30 °C on a rotating shaker (BR-43FL, TAITEC, Saitama, Japan) at 150 rpm. The cellulose products were collected by centrifugation (CR20GIII, Hitachi Koki, Tokyo, Japan) and dispersed into 1% w/v aqueous NaOH followed by incubation for 2 h at 70 °C with shaking, to lyse the remaining bacterial cells. After strong alkaline treatment, CM-NFBC was collected by centrifugation and washed with deionized water until it reached a pH of 6–7.

2.4. Preparation of DXR/CM-NFBC

DXR/CM-NFBC was prepared by mixing the DXR solution (2 mg/mL in Saline) with an equal volume of CM-NFBC (0.1, 0.2, 0.4, or 0.8% w/v). The concentration of DXR in the resultant DXR/CM-NFBC was adjusted to 1 mg/mL, and that of CM-NFBC was adjusted to 0.05, 0.1, 0.2, or 0.4% w/v.

2.5. Determining the Drug-Embedded Ratio in CM-NFBC

The DXR/CM-NFBC was centrifuged at 4 °C for 3 min at 15,000× g using a microcentrifuge system (KUBOTA, Osaka, Japan) to determine the retained amount of DXR. The supernatant was collected and diluted with MeOH (10 μL of supernatant with 100 μL of MeOH). The fluorescence intensity of the DXR in the supernatant samples was measured at an excitation wavelength of 470 nm and an emission wavelength of 590 nm using EnSpire™ (PerkinElmer, Waltham, MA, USA). The DXR concentration in the supernatant was determined from a preconstructed standard curve of DXR at various concentrations. The embedded ratio (%) of the DXR in the prepared DXR/CM-NFBC was then calculated using the following formula:

$$DXR\ embedded\ ratio\ (\%) = \left(\frac{DXR_{initial} - DXR_{supernatant}}{DXR_{initial}} \times 100 \right)$$

$DXR_{initial}$: a total initial amount of DXR
$DXR_{supernatant}$: an amount of free DXR in the supernatant

2.6. Release of DXR from DXR/CM-NFBC In Vitro

In vitro drug release from DXR/CM-NFBC formulations was evaluated in peritoneal wash fluid. In order to obtain peritoneal wash fluid, BALB/c mice were euthanized under deep anesthesia followed by intraperitoneal injection with 5 mL of saline to wash the peritoneal cavity, and then the peritoneal wash fluid was collected.

For in vitro release experiments, 500 μL of DXR/CM-NFBC was centrifuged at 4 °C for 3 min at 15,000× g, and the supernatant was discarded. Peritoneal wash fluid was then added onto the DXR/CM-NFBC pellet dropwise to a total volume of 1 mL. At selected time points post-incubation at 37 °C (0.5, 1, 2, 6 and 24 h), 100 μL of the supernatant was collected and thoroughly mixed with 300 μL of acetonitrile. After centrifugation (8000× g, 4 °C, 15 min), the supernatant (200 μL) was collected and mixed with 200 μL of water. DXR content was analyzed using a HPLC system (Shimadzu, Kyoto, Japan). The analysis was accomplished using a 4.6 × 150 mm C18 column (TSKgel ODS-100V) that was maintained at 35 °C with a mobile phase that consisted of acetonitrile and water (32:68, *v/v*; pH = 2.6). The flow rate was maintained at 1 mL/min, and the column effluent was monitored with an ultraviolet detector at 254 nm. The injection volume was 10 μL, and the retention time for DXR was 6.8 min. The DXR concentration was determined from a preconstructed calibration curve of DXR at various concentrations.

2.7. Preparation of Peritoneally Disseminated Gastric Cancer Xenograft Mouse Model

NCI-N87 human gastric carcinoma expressing firefly luciferase (NCI-N87-Luc, 128443) was purchased from Summit Pharmaceuticals International (Tokyo, Japan). The cells were cultured in RPMI-1640 medium (FUJIFILM Wako Pure Chemical Corporation) supplemented with 10% of heat-inactivated fetal bovine serum (COSMO BIO, Tokyo, Japan), 100 units/mL penicillin and 100 μg/mL streptomycin (ICN Biomedicals, Costa Mesa, CA, USA) in a 5% CO_2/air incubator at 37 °C.

To prepare a peritoneally metastatic gastric cancer mouse model, NCI-N87-Luc cells were intraperitoneally inoculated into BALB/c *nu/nu* mice (5×10^6 cells/mouse). Development and reduction of NCI-N87-Luc cells in the peritoneal cavity was confirmed using an in vivo imaging system (IVIS, Xenogen, Alameda, CA, USA) following an injection of *D*-luciferin potassium salt (100 μL of 7.5 mg/mL) under aesthesis with isoflurane inhalation. At 5 min post injection, bioluminescence derived from NCI-N87-Luc cells was observed with a CCD camera (30 s for the exposure time). The region of interest (ROI) in the bioluminescence was calculated and shown as bioluminescence intensity (BLI).

2.8. Evaluating the Tumor-Growth Inhibitory Effect of DXR/CM-NFBC on a Peritoneally Disseminated NCI-N87-Luc Xenograft Mouse Model

Peritoneally disseminated NCI-N87-Luc model mice were intraperitoneally injected with 4 doses of either free DXR or a formulation of DXR/CM-NFBC (0.5 mg DXR/kg/day) once a week from Day 8 post-tumor inoculation. At selected post-tumor inoculation timepoints (Days 7, 14, 21, 28, and 35), luciferase activity of the inoculated tumor cells in the peritoneal cavity was monitored with IVIS as aforementioned, and the body weights of the treated mice were recorded twice weekly. The antitumor effect was assessed in terms of both the bioluminescence intensity (BLI) and the percentage of tumor growth inhibition [TGI (%)]. The TGI was calculated using the following formula:

$$\text{TGI (\%)} = \left(1 - \frac{\text{BLI}_{treated}}{\text{BLI}_{control}}\right) \times 100$$

$\text{BLI}_{treated}$: A BLI in mouse treated with free DXR or DXR/CM-NFBC
$\text{BLI}_{control}$: A BLI in mouse non-treated

As a control, the model mice were intraperitoneally injected with 4 doses of CM-NFBC alone once a week from Day 8 post-tumor inoculation in amounts equal to the injections of DXR/CM-NFBC. At selected time points (Days 7, 13, 19, 25, and 31), luciferase activity of the peritoneally inoculated tumors was observed, and the BLI was calculated. At selected time points (Days 8, 14, 20, and 26), body weights of the mice were recorded.

2.9. Statistical Analysis

Statistical differences between the groups were evaluated by analysis of variance (ANOVA) with the Tukey post-hoc test, to explore the mean differences between pairs of groups, using Prism 8 software (GraphPad Software, San Diego, CA, USA). All values are reported as the mean ± S.D. The levels of significance were set at * $p < 0.05$, ** $p < 0.01$, and *** $p < 0.001$.

3. Results

3.1. Preparation of DXR/CM-NFBC Formulation

DXR is a widely used chemotherapeutic agent that was employed as an amphiphilic model drug for the preparation of a chemotherapeutic agent that could be embedded into CM-NFBC. The first step was to optimize the DXR/CM-NFBC formulation. To ensure efficient embedding, DXR was mixed with different concentrations of CM-NFBC (0.05, 0.1, 0.2, or 0.4% *w/v*). Figure 1A depicts the images of the prepared DXR/CM-NFBC after centrifugation. CM-NFBC was precipitated by centrifugation. At the lower CM-NFBC concentrations (0.05 or 0.1% *w/v*), strong red colors were observed in the supernatant, which indicated the presence of free/unembedded DXR. At relatively higher CM-NFBC concentrations (0.2 or 0.4% *w/v*), however, either few or no red colors were observed in the supernatant, which indicated that almost 100% of the added DXR had been embedded in the CM-NFBC. The DXR embedded ratios within each aliquot of DXR/CM-NFBC were calculated by measuring the fluorescent intensity of free DXR in the supernatant (Figure 1B). The DXR embedded ratios were directly proportional to CM-NFBC concentrations; the ratio was increased to 97.8% at 0.4% *w/v* CM-NFBC. The optimal mixture of DXR (2 mg/mL) and CM-NFBC (0.4% *w/v*) was adopted for further experiments.

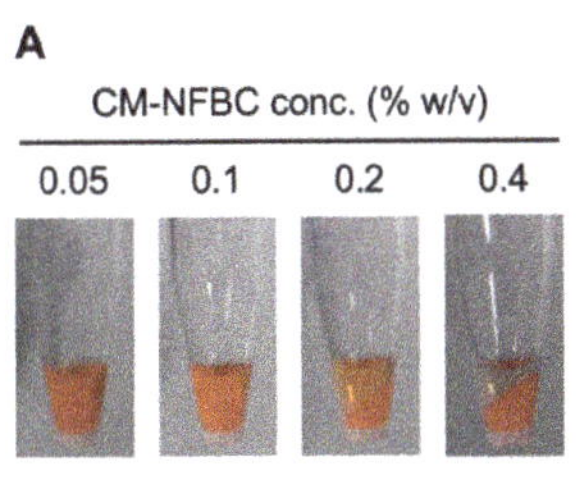

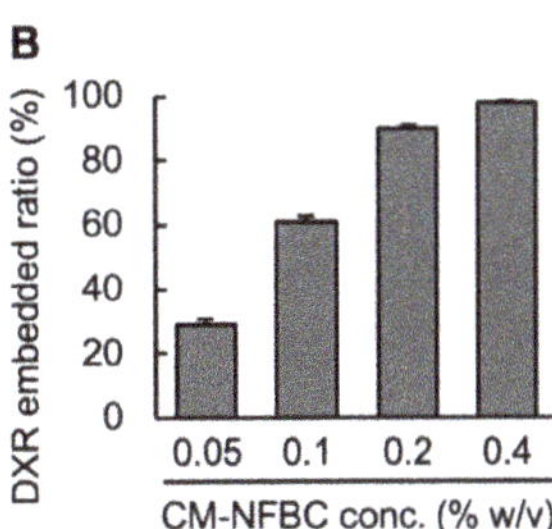

Figure 1. Preparation of DXR-embedding into CM-NFBC (DXR/CM-NFBC). DXR/CM-NFBC was prepared by mixing a DXR solution (2 mg/mL) with an equal volume of different concentrations of CM-NFBC (0.1, 0.2, 0.4, or 0.8% *w/v*). The concentration of DXR in the resultant DXR/CM-NFBC was adjusted to 1 mg/mL, and that of CM-NFBC was adjusted to 0.05, 0.1, 0.2, or 0.4% *w/v*. (**A**) DXR/CM-NFBC prepared with different concentrations of CM-NFBC. Photos were taken after centrifugation (15,000× *g*, 4 °C, 3 min). (**B**) Percentage of DXR embedded in the CM-NFBC. Data represent the mean ± SD (*n* = 6).

3.2. Release of Embedded DXR from the DXR/CM-NFBC Formulation in Peritoneal Fluid

One of the main purposes of this study was to apply the DXR/CM-NFBC formulation into the peritoneal cavity to treat peritoneally disseminated gastric cancer by intraperitoneal injection. Accordingly, the drug-release properties of DXR/CM-NFBC were evaluated via the incubation of DXR/CM-NFBC in the peritoneal wash fluid obtained from normal mice (Figure 2). Following 0.5 h of incubation, 26.6% of DXR was rapidly released from the formulation. Then, the release ratio reached a maximum level (79.5%) after 24 h of incubation. DXR/CM-NFBC appeared to achieve a sustained release of embedded DXR following intraperitoneal administration.

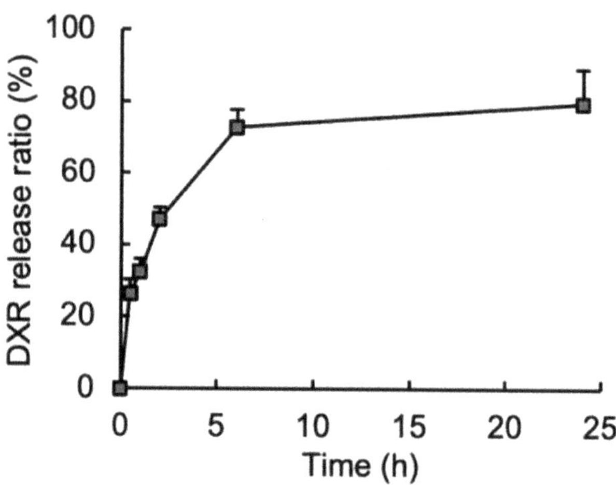

Figure 2. In vitro DXR release from DXR/CM-NFBC in peritoneal wash fluid. DXR/CM-NFBC (500 μL) was centrifuged and the supernatant was discarded. Peritoneal wash fluid was then added dropwise onto the DXR/CM-NFBC precipitates to a total volume of 1 mL. At selected time points post-incubation at 37 °C (0.5, 1, 2, 6 and 24 h), 100 μL of the supernatant was collected, and the DXR content was analyzed via HPLC. Data represent the mean ± SD ($n = 6$).

3.3. Growth Inhibitory Effect of an Intraperitoneally Injected DXR/CM-NFBC Formulation on Peritoneally Inoculated NCI-N87-Luc Tumors

To verify the tumor growth inhibitory effects, DXR/CM-NFBC was intraperitoneally injected into the mice in which NCI-N87-Luc cells had previously been peritoneally inoculated. The growth of inoculated cells was manifested by the increase in luciferase activity derived from NCI-N87-Luc cells before the injection of DXR/CM-NFBC. In the control group, luciferase activity gradually increased with time (Figure 3A,B), which indicated a steady growth of peritoneally inoculated cells. In both the free DXR and DXR/CM-NFBC groups (Figure 3A,B), luciferase activity was significantly decreased with time, suggesting that these treatments suppressed the number of inoculated cells in the peritoneal cavity. DXR/CM-NFBC treatment tended to produce a higher suppression effect than free DXR treatment. Treatment with drug-free CM-NFBC induced no tumor growth suppression (Figure 3C). To gain further insight, the TGI (%) was calculated from the values of BLI shown in Figure 3B (Table 1). These calculations confirmed that the treatment with either free DXR or DXR/CM-NFBC significantly decreased the BLI compared with that of the control group. In addition, treatment with DXR/CM-NFBC tended to induce a higher tumor growth inhibitory effect compared with treatment with free DXR. To confirm the systemic side effects of DXR/CM-NFBC, we monitored the body weight changes in peritoneally disseminated NCI-N87-Luc model mice following intraperitoneal treatment with DXR/CM-NFBC (Figure 3D). Treatment with free DXR clearly hampered efforts to increase the body weight of the treated mice, which indicated that the treatment with free DXR had induced the systemic side effects even with intraperitoneal treatment. In contrast to the control group, intraperitoneal treatment with DXR/CM-NFBC induced no significant body weight loss, and no body weight loss was observed in mice treated with drug-free CM-NFBC (Figure 3E).

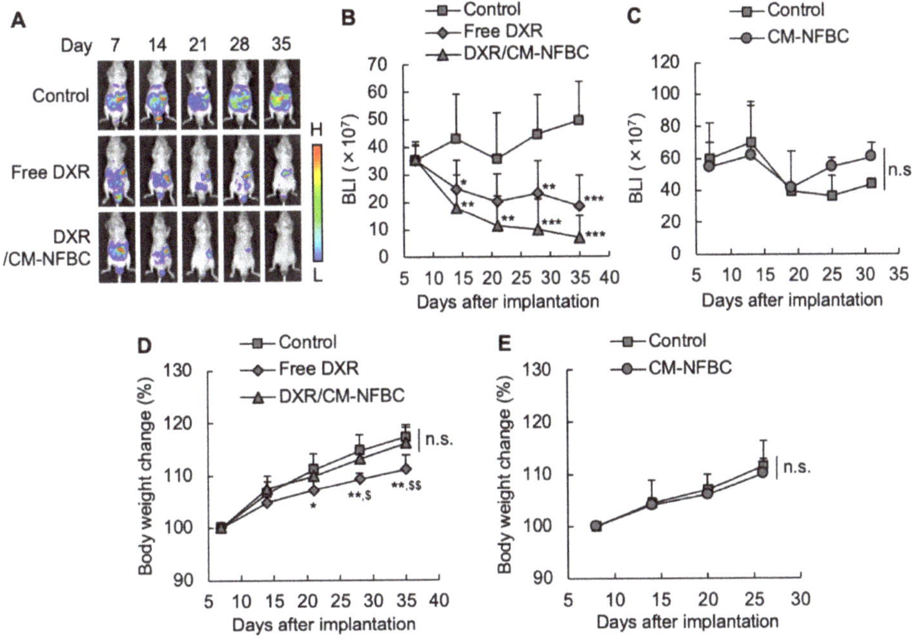

Figure 3. Growth inhibitory effect of an intraperitoneally injected DXR/CM-NFBC formulation on peritoneally inoculated NCI-N87-Luc tumors. Peritoneally disseminated gastric cancer (NCI-N87-Luc) xenograft model mice were intraperitoneally injected with four doses of either free DXR, the DXR/CM-NFBC formulation (0.5 mg DXR/kg/day), or CM-NFBC alone once a week from Day 8 post-tumor-inoculation. At selected time points post-tumor-inoculation, luciferase activity of the peritoneally metastatic tumors was monitored with IVIS. (**A**) Bioluminescence intensity (BLI) of peritoneally inoculated NCI-N87-Luc cells in each mouse treated with various formulations. Photos are typical examples from six mice. (**B**) The BLI of interest in the mice treated with various formulations. (**C**) The BLI of interest in the mice treated with CM-NFBC alone. (**D**) Body weight changes in the mice treated with various formulations. (**E**) Body weight changes in the mice treated with CM-NFBC alone. Data represent the mean ± SD (n = 6, * $p < 0.05$, ** $p < 0.01$, *** $p < 0.001$ vs. control, $ $p < 0.05$, $$ $p < 0.01$ vs. DXR/CM-NFBC, n.s.: no significance).

Table 1. Tumor growth inhibitory effects of the DXR/CM-NFBC formulation. (* $p < 0.05$, ** $p < 0.01$, *** $p < 0.001$ vs. control).

Treatment	Day 14		Day 21		Day 28		Day 35	
	BLI ($\times 10^7$)	TGI (%)	BLI ($\times 10^7$)	TGI (%)	BLI ($\times 10^7$)	TGI (%)	BLI ($\times 10^7$)	TGI (%)
Control	43.0 ± 16.1	- - - -	35.6 ± 16.8	- - - -	44.7 ± 14.2	- - - -	49.3 ± 14.3	- - - -
Free DXR	24.9 ± 10.3 *	42.1	20.3 ± 9.8	42.9	23.2 ± 11.6 **	48.0	18.6 ± 11.1 ***	62.4
DXR/CM-NFBC	18.0 ± 12.1 **	58.2	11.4 ± 11.0 **	67.9	10.0 ± 11.3 ***	77.7	7.0 ± 8.0 ***	85.8

4. Discussion

To date, BC has gained an ever-increasing amount of attention in the fields of biomedical and tissue engineering, which is a testament to qualities such as biocompatibility, flexibility, and mechanical strength [13,18,19]. However, only a few reports have mentioned the utility of BC as a drug-delivery vehicle. We recently reported that CM-NFBC, produced by culturing the strain identified as NEDO-01 under optimized conditions [15,16], could efficiently entrap the cytotoxic agent PTX, and the intraperitoneal administration of PTX-embedded CM-NFBC has shown superior therapeutic effects on a peritoneally disseminated gastric cancer xenograft mouse model [14]. To expand our work, in the present study we selected another anticancer agent, DXR, which is widely used to treat ovarian, bladder, breast, lung, thyroid, and stomach cancers [20–22]. CM-NFBC efficiently embedded DXR (Figure 1), and intraperitoneal administration of the formulation tended to suppress the peritoneal tumor growth and body weight loss, compared with that of the free DXR formulation, in a peritoneally disseminated gastric cancer xenograft mouse model (Figures 3 and 4). To the best of our knowledge, this is the first report to demonstrate the utility of CM-NFBC, not only for the delivery of a hydrophobic chemotherapeutic agent such as PTX, but also as a vehicle for an amphiphilic chemotherapeutic agent such as DXR.

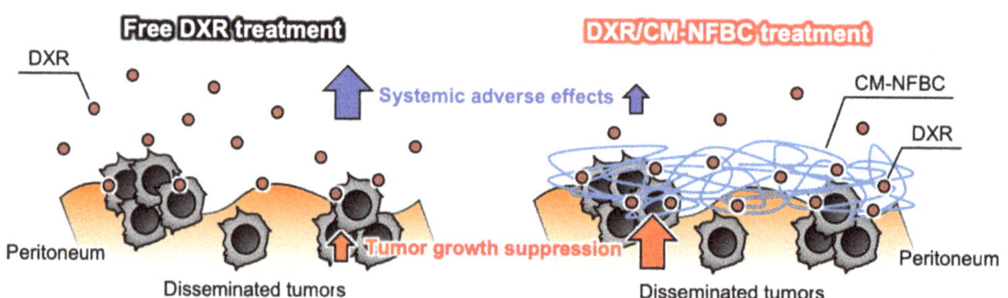

Figure 4. Schematic diagram of free DXR and DXR/CM-NFBC after intraperitoneal injection into a peritoneally disseminated tumor-xenograft mouse model.

Adequate drug release from a delivery vehicle at the target site of action is the key to gaining maximum therapeutic effect and to minimizing the side effects of embedded/encapsulated drugs [23]. DXR/CM-NFBC had efficiently released 60% of the embedded DXR into the peritoneal wash fluid within 6 h (Figure 2). The wash fluid mimics the in vivo environment of the target site of peritoneally disseminated gastric cancer xenograft mouse models. Therefore, the rapid and sustained release of DXR from the formulation in the present in vitro release study of a peritoneally disseminated cancer model (Figure 2) is a major indicator for the promise of a therapeutic effect with no body weight change (Figure 3).

The three-dimensional network structure of hydrogels such as CNFs tends to incorporate drugs and promote sustained release, which potentiates the antitumor effects and limits the undesirable side effects [24]. CM-NFBC is composed of hydrophilic cellulose networks with semi-micro sizes of pores and is characterized by relatively uniform nanofibers with a diameter of ca. 20 nm [15]. Due to such a unique structure and characteristics, DXR was efficiently trapped by the hydrophilic cellulose network when vigorously mixed with CM-NFBC. Furthermore, the -COOH groups in the hydrogel were ionized into carboxyl ions (-COO$^-$) in a neutral condition that expanded the network structure via electrostatic repulsion, which allowed the absorption of water into the hydrogel [25,26]. CM-NFBC is composed of nanofibrillated cellulose, and when modified with CMC on the surfaces of the nanofibers, this formulation contains -COOH groups owing to the high frequency of mixing [15,16]. Following the intraperitoneal injection of DXR/CM-NFBC, these -COOH groups would likely expand the cellulose network of NFBC in the peritoneal cavity and absorb water, which should enhance the release of DXR from the DXR/CM-NFBC formulation.

Anthracycline anticancer drugs, such as DXR, daunorubicin, and epirubicin, are an important class of chemotherapeutic agents for the treatment of a wide variety of solid tumors. Notably, DXR is a key chemotherapeutic drug for cancer treatment, although its clinical use is limited by acute and chronic adverse events. In a retrospective analysis of more than 4000 patients receiving DXR, 2.2% developed clinical signs and symptoms of congestive heart failure [27]. The toxicity that DXR exhibits in cardiomyocytes is related to free radical formation caused by the metabolism of DXR after accumulation in the heart [28], which means that a transfer of DXR into the circulation and its subsequent accumulation in the heart would likely induce severe cardiotoxicity. Indeed, the intraperitoneal injection of free DXR caused substantial body weight loss in a mouse model (Figure 3D) probably due to the quick distribution of DXR into circulation. By contrast, an intraperitoneal injection of DXR/CM-NFBC did not decrease the body weight of the mice compared with the non-treatment control mice (Figure 3D), which suggests a relatively lower concentration of DXR in the peritoneal cavity and blood circulation due to the sustained release of DXR from the DXR/CM-NFBC formulation (Figure 2).

In our previous study, when PTX, a hydrophobic anticancer drug, was embedded in CM-NFBC it was barely released from the formulation in the peritoneal wash fluid; less than 3% of embedded PTX was released even after 24 h of incubation [14]. Taken together with the results shown in Figures 1 and 2, it appears that the physicochemical properties of drugs mainly govern their levels of embedding into CM-NFBC as well as their release from the formulation. Both amphiphilic drugs such as DXR and hydrophobic drugs such as PTX are strongly embedded into CM-NFBC. Amphiphilic drugs such as DXR might be quickly released from the formulation, while hydrophobic drugs such as PTX might be retained within the formulation.

The in vivo imaging study showed that DXR/CM-NFBC successfully tended to increase the suppressive effect of DXR on tumor growth in a peritoneally disseminated gastric cancer xenograft mouse model compared with the free DXR formulation (Figure 3A,B). Intraperitoneal chemotherapy has recently been introduced as a smart strategy to treat peritoneally disseminated cancers, because intraperitoneal chemotherapy can directly expose metastatic tumors to the drugs of interest [29,30]. Nevertheless, intraperitoneally injected anticancer drugs, notably hydrophilic/amphiphilic ones such as DXR and cisplatin, tend to be rapidly cleared from the peritoneal cavity via the circulatory system [31], and thus no effective treatment with intraperitoneal chemotherapy has yet improved the survival rates of patients with peritoneal dissemination of cancers. Our drug/CM-NFBC formulation could represent a suitable pharmaceutical formulation for clinical settings in achieving the long peritoneal retention of hydrophilic/amphiphilic agents following intraperitoneal injection.

5. Conclusions

A novel formulation, DXR/CM-NFBC, was successfully prepared by embedding the amphiphilic anticancer drug, DXR. This formulation showed a somewhat higher therapeutic efficacy in a peritoneally metastatic gastric cancer xenograft mouse model compared with the use of free DXR. We collectively demonstrated that our NFBC is a potentially powerful drug delivery vehicle for various anticancer agents used in the treatment of peritoneally disseminated cancers via intraperitoneal injection.

Author Contributions: Conceptualization, H.A. and T.I.; methodology, H.A., T.M. (Takashi Mochizuki) and S.A.; software, H.A. and K.F.; validation, T.M. (Takashi Mochizuki) and S.A.; formal analysis, H.A.; investigation, T.M. (Takashi Mochizuki) and S.A.; resources, K.T., T.M. (Tokuo Matsushima) and T.K.; data curation, H.A., T.S. and Y.I.; writing—original draft preparation, H.A. and A.S.A.L.; writing—review and editing, A.S.A.L. and T.I.; visualization, H.A.; supervision, T.I.; project administration, H.A.; funding acquisition, K.T. and T.I. All authors have read and agreed to the published version of the manuscript.

Funding: This study was in part supported by Grant-in-Aid for Scientific Research (B) (19H02549) from the Japan Society for the Promotion of Science, and a research program for the development of an intelligent Tokushima artificial exosome (iTEX) from Tokushima University.

Institutional Review Board Statement: The study was conducted according to the guidelines of the Declaration of Helsinki, and approved by the Animal and Ethics Review Committee of Tokushima University (the approval code: T2019-47).

Informed Consent Statement: Not applicable.

Data Availability Statement: The data presented in this study are available on request from the corresponding author.

Conflicts of Interest: The authors declare no conflict of interest. The funders had no role in the design of the study; in the collection, analyses, or interpretation of data; in the writing of the manuscript, or in the decision to publish the results.

References

1. Siddique, S.; Chow, J.C.L. Application of Nanomaterials in Biomedical Imaging and Cancer Therapy. *Nanomaterials* **2020**, *10*, 1700. [CrossRef]
2. Somerville, C. Cellulose synthesis in higher plants. *Annu. Rev. Cell Dev. Biol.* **2006**, *22*, 53–78. [CrossRef]
3. Doblin, M.S.; Kurek, I.; Jacob-Wilk, D.; Delmer, D.P. Cellulose biosynthesis in plants: From genes to rosettes. *Plant Cell Physiol.* **2002**, *43*, 1407–1420. [CrossRef] [PubMed]
4. Ross, P.; Mayer, R.; Benziman, M. Cellulose biosynthesis and function in bacteria. *Microbiol. Rev.* **1991**, *55*, 35–58. [CrossRef] [PubMed]
5. Jozala, A.F.; de Lencastre-Novaes, L.C.; Lopes, A.M.; de Carvalho Santos-Ebinuma, V.; Mazzola, P.G.; Pessoa, A., Jr.; Grotto, D.; Gerenutti, M.; Chaud, M.V. Bacterial nanocellulose production and application: A 10-year overview. *Appl. Microbiol. Biotechnol.* **2016**, *100*, 2063–2072. [CrossRef] [PubMed]
6. Keshk, S.M. Vitamin C enhances bacterial cellulose production in Gluconacetobacter xylinus. *Carbohydr. Polym.* **2014**, *99*, 98–100. [CrossRef]
7. Saxena, I.M.; Brown, R.M., Jr. Cellulose biosynthesis: Current views and evolving concepts. *Ann. Bot.* **2005**, *96*, 9–21. [CrossRef]
8. Haigler, C.H.; White, A.R.; Brown, R.M., Jr.; Cooper, K.M. Alteration of in vivo cellulose ribbon assembly by carboxymethylcellulose and other cellulose derivatives. *J. Cell Biol.* **1982**, *94*, 64–69. [CrossRef]
9. Lin, D.; Lopez-Sanchez, P.; Li, R.; Li, Z. Production of bacterial cellulose by Gluconacetobacter hansenii CGMCC 3917 using only waste beer yeast as nutrient source. *Bioresour. Technol.* **2014**, *151*, 113–119. [CrossRef] [PubMed]
10. Petersen, N.; Gatenholm, P. Bacterial cellulose-based materials and medical devices: Current state and perspectives. *Appl. Microbiol. Biotechnol.* **2011**, *91*, 1277–1286. [CrossRef] [PubMed]
11. Bhandari, J.; Mishra, H.; Mishra, P.K.; Wimmer, R.; Ahmad, F.J.; Talegaonkar, S. Cellulose nanofiber aerogel as a promising biomaterial for customized oral drug delivery. *Int. J. Nanomed.* **2017**, *12*, 2021–2031. [CrossRef]
12. Khoshnevisan, K.; Maleki, H.; Samadian, H.; Shahsavari, S.; Sarrafzadeh, M.H.; Larijani, B.; Dorkoosh, F.A.; Haghpanah, V.; Khorramizadeh, M.R. Cellulose acetate electrospun nanofibers for drug delivery systems: Applications and recent advances. *Carbohydr. Polym.* **2018**, *198*, 131–141. [CrossRef] [PubMed]
13. Lobmann, K.; Svagan, A.J. Cellulose nanofibers as excipient for the delivery of poorly soluble drugs. *Int. J. Pharm.* **2017**, *533*, 285–297. [CrossRef]
14. Akagi, S.; Ando, H.; Fujita, K.; Shimizu, T.; Ishima, Y.; Tajima, K.; Matsushima, T.; Kusano, T.; Ishida, T. Therapeutic efficacy of a paclitaxel-loaded nanofibrillated bacterial cellulose (PTX/NFBC) formulation in a peritoneally disseminated gastric cancer xenograft model. *Int. J. Biol. Macromol.* **2021**, *174*, 494–501. [CrossRef] [PubMed]
15. Kose, R.; Sunagawa, N.; Yoshida, M.; Tajima, K. One-step production of nanofibrillated bacterial cellulose (NFBC) from waste glycerol using Gluconacetobacter intermedius NEDO-01. *Cellulose* **2013**, *20*, 2971–2979. [CrossRef]
16. Tajima, K.; Kusumoto, R.; Kose, R.; Kono, H.; Matsushima, T.; Isono, T.; Yamamoto, T.; Satoh, T. One-Step Production of Amphiphilic Nanofibrillated Cellulose Using a Cellulose-Producing Bacterium. *Biomacromolecules* **2017**, *18*, 3432–3438. [CrossRef]
17. Saito, T.; Nishiyama, Y.; Putaux, J.L.; Vignon, M.; Isogai, A. Homogeneous suspensions of individualized microfibrils from TEMPO-catalyzed oxidation of native cellulose. *Biomacromolecules* **2006**, *7*, 1687–1691. [CrossRef] [PubMed]
18. Portela, R.; Leal, C.R.; Almeida, P.L.; Sobral, R.G. Bacterial cellulose: A versatile biopolymer for wound dressing applications. *Microb. Biotechnol.* **2019**, *12*, 586–610. [CrossRef] [PubMed]
19. Sharma, A.; Thakur, M.; Bhattacharya, M.; Mandal, T.; Goswami, S. Commercial application of cellulose nano-composites—A review. *Biotechnol. Rep.* **2019**, *21*, e00316. [CrossRef] [PubMed]
20. Valero, V.; Perez, E.; Dieras, V. Doxorubicin and taxane combination regimens for metastatic breast cancer: Focus on cardiac effects. *Semin. Oncol.* **2001**, *28*, 15–23. [CrossRef] [PubMed]
21. Roth, B.J.; Bajorin, D.F. Advanced bladder cancer: The need to identify new agents in the post-M-VAC (methotrexate, vinblastine, doxorubicin and cisplatin) world. *J. Urol.* **1995**, *153*, 894–900. [CrossRef]
22. Aisner, J.; Whitacre, M.; Abrams, J.; Propert, K. Doxorubicin, cyclophosphamide, etoposide and platinum, doxorubicin, cyclophosphamide and etoposide for small-cell carcinoma of the lung. *Semin. Oncol.* **1986**, *13*, 54–62.
23. Zhu, Q.; Qi, H.; Long, Z.; Liu, S.; Huang, Z.; Zhang, J.; Wang, C.; Dong, L. Extracellular control of intracellular drug release for enhanced safety of anti-cancer chemotherapy. *Sci. Rep.* **2016**, *6*, 28596. [CrossRef]
24. Wei, W.; Li, H.; Yin, C.; Tang, F. Research progress in the application of in situ hydrogel system in tumor treatment. *Drug Deliv.* **2020**, *27*, 460–468. [CrossRef] [PubMed]
25. Sahoo, B.; Devi, K.S.; Banerjee, R.; Maiti, T.K.; Pramanik, P.; Dhara, D. Thermal and pH responsive polymer-tethered multifunctional magnetic nanoparticles for targeted delivery of anticancer drug. *ACS Appl. Mater. Interfaces* **2013**, *5*, 3884–3893. [CrossRef] [PubMed]

26. Rizwan, M.; Yahya, R.; Hassan, A.; Yar, M.; Azzahari, A.D.; Selvanathan, V.; Sonsudin, F.; Abouloula, C.N. pH Sensitive Hydrogels in Drug Delivery: Brief History, Properties, Swelling, and Release Mechanism, Material Selection and Applications. *Polymers* **2017**, *9*, 137. [CrossRef] [PubMed]
27. Von Hoff, D.D.; Layard, M.W.; Basa, P.; Davis, H.L., Jr.; Von Hoff, A.L.; Rozencweig, M.; Muggia, F.M. Risk factors for doxorubicin-induced congestive heart failure. *Ann. Intern. Med.* **1979**, *91*, 710–717. [CrossRef] [PubMed]
28. Davies, K.J.; Doroshow, J.H. Redox cycling of anthracyclines by cardiac mitochondria. I. Anthracycline radical formation by NADH dehydrogenase. *J. Biol. Chem.* **1986**, *261*, 3060–3067. [CrossRef]
29. Kono, K.; Yong, W.P.; Okayama, H.; Shabbir, A.; Momma, T.; Ohki, S.; Takenoshita, S.; So, J. Intraperitoneal chemotherapy for gastric cancer with peritoneal disease: Experience from Singapore and Japan. *Gastric Cancer* **2017**, *20*, 122–127. [CrossRef]
30. Kobayashi, D.; Kodera, Y. Intraperitoneal chemotherapy for gastric cancer with peritoneal metastasis. *Gastric Cancer* **2017**, *20*, 111–121. [CrossRef] [PubMed]
31. Chambers, S.K.; Chow, H.H.; Janicek, M.F.; Cragun, J.M.; Hatch, K.D.; Cui, H.; Laughren, C.; Clouser, M.C.; Cohen, J.L.; Wright, H.M.; et al. Phase I trial of intraperitoneal pemetrexed, cisplatin, and paclitaxel in optimally debulked ovarian cancer. *Clin. Cancer Res.* **2012**, *18*, 2668–2678. [CrossRef] [PubMed]

MDPI

St. Alban-Anlage 66

4052 Basel

Switzerland

Tel. +41 61 683 77 34

Fax +41 61 302 89 18

www.mdpi.com

Nanomaterials Editorial Office

E-mail: nanomaterials@mdpi.com

www.mdpi.com/journal/nanomaterials